GLIMPSE ON
GENERAL AGRICULTURE

About the Author

P. Laxman Rao is pursuing his Ph.D. from Professor Jayashankar Telangana State Agricultural University, Rajendranagar, Hyderabad. He is a M.Sc (Ag.) in Soil Science and Agricultural Chemistry (2015) from PJTSAU, Hyderabad and a graduate from Acharya N.G. Ranga Agriculture University, Hyderabad (2013). He worked as a Senior Research Fellow on the project 'CRP - CA at CRIDA'.

I. Thirupathi, is pursuing his Ph.D. from Professor Jayashankar Telangana State Agricultural University, Rajendranagar, Hyderabad. He completed his M.Sc (Ag.) in Agronomy from Acharya N.G. Ranga Agriculture University, Hyderabad (2014) and his B.Sc (Ag) from Acharya N.G. Ranga Agriculture University, Hyderabad (2012).

Purma Sadvi, is a Ph.D. Research Scholar at Mahathma Phule Krishi Vidyapeeth, Rahuri, Maharastra. She pursued her M.Sc (Ag.) in Agricultural Extension (2015) from Professor Jayashankar Telangana State Agricultural University, Hyderabad and is a graduate from Acharya N.G. Ranga Agriculture University, Hyderabad (2013).

GLIMPSE ON

GENERAL AGRICULTURE

(For ICAR-JRF, SRF, NET and ASRB Prelims)

Compiled by

P. Laxman Rao

I. Thirupathi

Purma Sadvi

SCIENTIFIC PUBLISHERS

Published by
SCIENTIFIC PUBLISHERS (INDIA)

Jodhpur –

5 A, New Pali Road
P.O. Box 91
Jodhpur - 342 001 INDIA

Delhi –

4806/24, Ansari Road
Daryaganj
New Delhi - 110 002 INDIA

ISBN: 978-93-86652-45-4

Visit the Scientific Publishers (India) website at
http://www.scientificpub.com
Printed in India

Preface

Agriculture is a broad subject. It covers all the basic subjects of science. Therefore students have to pass through all the agriculture related subjects with elementary courses during under graduation. After passing graduation, higher degrees are done in specialized field of Agriculture so there is no need to read all Agriculture subjects in M. Sc. and Ph. D. But for the preparation of various agriculture competition examinations students have to read all the basic books of Agriculture to cover syllabus. That time most of students don't have all the necessary books and too much time to read them.

Therefore to overcome these problems the authors tried their best to write this book through reading various books and other sources of Agriculture to cover brief and best information of subjects (Agronomy, Genetics and Plant Breeding, Soil Science and Agricultural Chemistry, Agricultural Microbiology, Physiology, Agricultural Entomology, Plant Pathology, Horticulture, Agricultural Economics, Agricultural Extension And Agricultural Statistics) in one book.

To covering these ten subjects of Agriculture in one book is very hard task so some mistake may be left, therefore your suggestions are invited to further improvement of this book. But authors would like to say that students have to treat this book as a competitive book not a text book for various degree courses.

I hope this book would be very useful for students who appear in various Agricultural competition examinations.

Dr. M. RAJESHWAR NAIK

Associate Professor (SS & AC)
College of Agriculture, PJTSAU,
Rajendranagar, Hyderabad-30

Chapter Contributors

P. Laxman Rao
Ph. D. Research Scholar
Dept. of Soil Science and Agricultural Chemistry
PJTSAU, Rajendranagar, Hyderabad,
Chapter contributed: Soil Science and Agricultural Chemistry

I. Thirupathi
Ph. D. Research Scholar
Dept. of Agronomy
PJTSAU, Rajendranagar, Hyderabad.
Chapter contributed: Agronomy and Horticulture

Purma Sadvi
M. Sc (Agriculture)
Dept. of Agricultural Extension
PJTSAU, Rajendranagar, Hyderabad.
Chapter contributed: Agricultural Extension

Gade Priyanka
Pn. D. Research Scholar
Dept. of Genetics and Plant Breeding
PJTSAU, Rajendranagar, Hyderabad.
Chapter contributed: Genetics and Plant Breeding

L. Mahesh
M.Sc. (Agriculture)
Dept. of Agricultural Extension
PJTSAU, Rajendranagar, Hyderabad.
Chapters contributed: Agricultural Extension, and Agricultural Economics

B. Madavi
M.Sc. (Agriculture)
Dept. of Agronomy
PJTSAU, Rajendranagar, Hyderabad.
Chapters contributed: Agronomy and Agricultural statistics

E. Govardhan
M.Sc. (Agriculture)
Dept. of Soil Science and Agricultural Chemistry
PJTSAU, Rajendranagar, Hyderabad
Chapters contributed: Agricultural Microbiology and Physiology

B. Thirupam Reddy
Ph.D. Research Scholar
Dept. of Agricultural Entomology
UAS, Dharwada, Karnataka.
Chapters contributed: Agricultural Entomology, and Plant Pathology

Contents

CHAPTER 1

Agronomy

1. Embryo development without fertilization? **Parthenocarpy**
2. Exhanthema Disease caused by **Cu.**
3. First DG of ICAR **Habibulla 1965.**
4. First state to adopt panchayut raj system (2 tier) **Rajasthan.**
5. Food corporation of India established in? **1965.**
6. Downey mildew of grape caused by? ***Plasmopara viticola***
7. Sand % very low in ? **Silty loam.**
8. IGRRI located at ? **Jhansi.**
9. Test weight of phalaris minor **0.5 gram.**
10. Particle Density of soil always **Greater than B.D.**
11. Zero competition seen in **Parallel cropping.**
12. 2.4-D very sensitive to ? **Cotton.**
13. Grassy shoot in sugarcane are to **MLOs.**
14. Gundhi bug attacks at **Milking stage.**
15. Highest water use efficiency **CAM > C_4 > C_3.**
16. Ian wilmut associated with **Creation of dolly.**
17. Opoque-2 [QPM] rich in_______? **Methionine, Thymine and lysine.**
18. First transgenic plant developed in world? **Tobacco.**
19. False smut is a disease of ? **Rice.**
20. $K = K_2O$ X _______ **0.83.**
21. Pre emergent herbicide? **Fluchloralin**
22. Conservation tillage minimum _______% of residue? **45%**
23. Fallowing which tree Sps not yielding timber? **Subabul.**

24. Protein content of soybean? **40 – 42%**
25. The word Agronomy derived from? **Greek**
26. Black soils are also called as? **Verti soils**
27. 1 bale of jute? **180 kg**
28. Unit of WUE ? **Kg/ha-mm**
29. Alkaloid present in castor? **Ricinin**
30. Active ingredient of neem which of used as insecticide? **Azadirachtin**
31. Active ingredient of lemon grass ? **Citronella**
32. C:N ratio of idea soil of **10 to 12 : 1**
33. Fertility variation in two direction then which design as preferable? **LSD**
34. Stomatal movement of regulated by? **K**
35. Cereal which of low in protein content ? **Rice**
36. VAM helps **P mobilization**
37. Economic yield of 40 kg. Biological yield 60 kg. what is HI? **60%**
38. Triticale is cross between **Wheat × Rye**
39. CIMMYT located at ? **Mexico**
40. Anaerobic symbiotic N_2 Fixer? **Clostridium**
41. Expand FAO? **Food and agriculture organization**
42. Head of sunflower is called? **Capitula**
43. Organic matter = **0.C × 1.1724**
44. Word Remote sensing coined by **Mather.**
45. Sulfosulfuron dosage for wheat **25 g/ha.**
46. Correlation coefficient lies between **–1 to +1.**
47. Self pollination plant is? **Rice**
48. Green manure crop suitable for water logged situation? **Dhaianch**
49. Stress harmone ______? **ABA**
50. At PWP the soil moisture tention at **31 bar.**
51. Indian society of agronomy established in **1955.**
52. Lab to land programme started in **1st June 1979.**
53. Minimum sample size required for chi square test **12.**
54. Pest occurs in same area year after year? **Epidemic**
55. Tea yellow disease occurs due to deficiency of ? **S**
56. Transgenic crop having maximum area in world. **Soybean**

57. Non selective, contact herbicide ? **Paraquat**
58. Non selective translocated herbicide ? **Glyphosate**
59. Fragmentation of land takes place throw? **Gully erosion**
60. Available water held at tension **-1/3 to -15 bar**
61. Primary resources for agriculture **Land.**
62. Water logged soil, availability of fallowing increases? **Fe, Mn, P.**
63. EC of saline soil **> 4 dS/m**
64. Gene responsible for dwarfness of wheat ? **Dee geo woo gen**
65. Water suckers related to which crop ? **Sugarcane**
66. T_z or tetrazolium test used for? **Seed viability**
67. 17th Essential plant nutrient? **Nickel**
68. AGMARK headquatres located at ? **Nagpur**
69. Micronutrient required for N Fixation ? **Mo**
70. Fertilizer suitable for flooded paddy ? **Ammonium nitrate**
71. Rice – Wheat – Rice – Wheat, rotational intensity (2 year) **200%**
72. Retting temperature for jute? **34°C**
73. Photo respiration cycle is also called as ? **Krebs cycle**
74. Which is not true regarding urea ? **Slowly soluble fertilizer**
75. From which part, Jute fiber obtaining? **Paranchyma tissue (stem)**
76. Lapse rate ? **6.5°C/km**
77. Popping in Ground nut due to deficiency of ? **Ca**
78. Longest day -------? **June 21**
79. Water which held around the soil particles as tight, thin film is? **Hygroscopic water**
80. Runoff coefficient of higher in ? **Black soil**
81. Potato root is called? **Stolon**
82. Law of manimum given by ? **Leibig**
83. Pollen sterility caused due to deficiency of ? **B**
84. Isoproturon belongs to which class of herbicide **Ureas**
85. Fallowing which crop is not a short day plant? **Wheat**
86. Japonica var of rice mainly grown in ? **Japan**
87. Flint corn is ***Zea mays* indurata**
88. Redgram germination type? **Epigeal**

89. Climatic classification suitable to agricultural purpose given by? **Koppen**
90. Meteorological subdivisions in India **36**
91. "At constant temperature the density of gas varies directly to pressure" the law of **Boyle's law**
92. 1 Nautical mile is? **1.8 km/hr**
93. Critical limit for nitrate in water? **10 PPM**
94. Water requirement ? **DR+ER+S**
95. Observation drawn from sample is? **Sampling**
96. Cross inoculation for soybean? ***R. Japonicum***
97. CO_2 Fixer in C_3 plants ? **Rubisco**
98. In rice reproductive stage stars at? **PI stage**
99. Critical stage for irrigation in wheat? **CRI**
100. Red colour of soil of due? **Iron oxides**
101. Commonly cultivation season for wheat in india? **Winter**
102. Spear mint scientific name? **Mentha spicate**
103. Dormancy of seed potato is broken by? **GA**
104. Average land holding of Indian farmer? **1.15 ha**
105. In saturated soil which law is applicable? **Darcys law**
106. At high PH which micronutrient is available? **Mo**
107. IFOAM located at? **Benn, Germany**
108. Urea is? **Amide**
109. Double zero is variety of? **Mustard**
110. Cell theory given by ? **Schleiden & Schwan**
111. Gene responsible for N Fixation? **Nif gene**
112. Respiratory quotient means (RQ) **Volume of CO_2 evolved / volume of O_2 released.**
113. RQ for organic acid? **1.33**
114. Enzymes are basically? **Protein**
115. Calcium is Important constituent of ? **Cell wall**
116. Enzyme that act on fat ? **Lipase**
117. Heart rot of sugar beet is due to deficiency of ? **B**
118. 'Sand drawn' of tobacco due to? **Mg**
119. Natural inhibitor present in seed ? **ABA**

120. Pollen shedder found in which crop? **Maize**
121. Basic unit of soil class? **Soil order**
122. Maximum density of water at. **4°C**
123. Father of soil testing ? **M.L. Trough**
124. Freckling in sugarcane due to deficiency of ? **Mn**
125. Highest pore space found in ? **Black soil**
126. High field capacity found in ? **Black soil**
127. Crop more tolerant to salinity ? **Barley**
128. Normally C:N:S ratio of soil? **100:10:1**
129. Critical limit for available sulpher in soil? **20 ppm.**
130. Long term persistent herbicide? **Pendimethalin**
131. Lucknow-49 of popular vanity of ? **Guava**
132. Purple blotch is problem in? **Onion**
133. Acid present in grape? **Malic acid**
134. 'Snowball 16' is a varity of ? **Cauliflower**
135. Largest chilli growing state? **Ap**
136. The total area under wheat in India (m.ha) **30.**
137. The state which produces largest quantity of potato in India is? **UP**
138. The largest hectarage and production amongst the field crops in India is associated with? **Rice**
139. Which Agricultural University in India was established first? **GBPUAT, Pant Nagar, UP.**
140. A farmer has provision of one irrigation for wheat crop. At what stage he should irrigate the crop? **CRI**
141. Nursery area required to provide seedling of rice to transplant one hectare field is? **0.10 ha**
142. Which fertilizer is fire hazardous? **Ammonium nitrate**
143. Which crops contribute the maximum to the production of vegetable oil in the country? **Groundnut**
144. The Central Institute for Post – Harvest Engineering and Technology (CIPHET) is located in **Ludhiana.**
145. What has the highest irrigation efficiency? **Drip**
146. How many Deemed Universities are part of the Indian Council of Agril. Research **Four.**

147. Who was the first Director General of the recognized ICAR in 1965? **Dr. B.P. Pal**
148. The unit of rural society is **Village.**
149. 'Khaira' disease of rice is due to **Zinc deficiency.**
150. The first district level extension programme launched in independent India was in **Amritsar (Punjab).**
151. If the fertilizer is to be applied at the rate of 100 kg per hectare, how much will be needed for a plot of 50 sq. m **0.50 kg.**
152. Among the following, which is the monophagous insect pest **Yellow stem borer of rice.**
153. A lady bird beetle is **A predator.**
154. In a cell, DNA synthesis takes place in **Interphase.**
155. The standard error a simple mean is given by **SE = $s/\sqrt{n}$**
156. 'Phyllody' is an important disease in **Sesamum.**
157. Which among the following is most salt tolerant fruit crop **Date palm.**
158. Photorespiration involves **Oxidative photosynthetic carbon cycle.**
159. Which one of the following river basin has maximum utilizable flow **Ganga.**
160. A line drawn on a map connecting points of equal rainfall is **Isohyte.**
161. The seed of a crop variety produced by the breeder which is small in quantity is said to be **Nucleus seed.**
162. Dapog seedlings would be ready for transplanting within: **11-14 days.**
163. Who is the present Chairman of Protection of Plant Varieties and Farmers' Right Authority (PPV&FRA)? **Dr. R.R. Hanchinal.**
164. Dog ridge is **Salt tolerant rootstocks of mango.**
165. Photosynthetically active radiation (PAR) is measured by **Quantum sensor.**
166. The base temperature of rice for calculating GDD is **10°C.**
167. A film-forming anti-transpirant used indryland agriculture is **Mobileaf.**
168. One Baule unit of P (lb P2O5/acre) corresponds to **45.**
169. The value of 'K' (lb/acre) used in the Wilcox's'Inverse-yield N' law is **418.**
170. Darcy's law is valid for **Only saturated flow.**

171. Which part of the leaf is present in wheat but not in the grassy weeds such as *Phalarisminor*, wild oats? **Ligule.**
172. The most common growth limiting factor in acid soil is **Aluminium.**
173. Growing degree day concept is useful inforecasting of **Crop harvest dates.**
174. Pungency in mustard oil is due to **Glucosinolates.**
175. Percent calcium content in single superphosphate is **12.**
176. *Eruca sativa* is a **Vegetable.**
177. The ratio between the irrigated area and the volume of water used is known as **Duty of water.**
178. The sensitivity of tensiometer is only up to **0.85 atm.**
179. Movement of P and K from soil to the root surface take place due to **Mass flow.**
180. Which design is suitable when fertility gradient is in two directions? **Split-plot design.**
181. What is the chemical formula of slaked lime? **$Ca(OH)_2$.**
182. Which degradation plays a major role in the persistence and behaviour of herbicides in soil? **Chemical.**
183. The triazine group of herbicides act by **Inhibiting root and shoot growth.**
184. The first mustard variety developed by somatic hybridization is **Pusa Jaikisa.**
185. Deflocculation is due to **Na.**
186. Which year is called International year of soil? **2015.**
187. *Chenopodium album* is the scientific name of **Bathua.**
188. Contribution of flag leaf in photosynthates is about **52%.**
189. Soil aggregation is determined by using **Yoders apparatus.**
190. RMO-40 is a Variety of **mothbean.**
191. Most critical stage for irrigation and fertilizer application in cereals is **Panicle Initiation.**
192. Which type of maize is early maturing one? **Pop corn**
193. Which crop is have the highest protein content? **Soybean**
194. Optimum seed rate for sowing Chickpea **75-100 kg ha^{-1}.**
195. Which of the operation is specially followed in chickpea **Nipping.**

196. Highest area and production of gram is in **Madhya Pradesh**.
197. Minimum per capita pulses/gm/day recommended by ICMR is **37**.
198. Which crop has the highest cultivated area in the world **Wheat**.
199. White bud of maize symptom is due to the deficiency of **Zn**.
200. Seed rate of barley is **75-80 Kg/ha**.
201. Which among the crop is having the highest oil content ? **Sesamum**
202. Optimum seed rate for sowing Rice. **125-150 kg ha^{-1}**
203. Which of the operation is specially followed in Castor ? **Nipping of auxillary racemes.**
204. Highest wheat producing state in India **Uttar Pradesh**.
205. KBC-2 is a variety of **Cowpea**.
206. Which is an example of Nitro positive crop? **Maize**
207. Denitrification losses of nitrogen in paddy field can be prevented by incorporating ammonical nitrogen in **Reduced zone**.
208. Well fitted inter cropping system for black soil during *Rabi* season is **Wheat + safflower (6:2)**.
209. The multicut fodder variety of Sorghum is **Meethisudan**.
210. Sorghum poisoning is caused by the presence of **Prussic acid**.
211. J-Set-3 is a fodder variety of **Sorghum**
212. Seed rate of wheat under irrigated conditions is **100-125 kg/ha**.
213. In maize, appearance of 'dead heart' symptom is an indication of **Stem borer attack**.
214. Highest production of ragi in India is in **Karnataka**.
215. *Khaira* disease of rice is caused due to **Zn Deficiency**.
216. What is easily digestable ? **Black gram**
217. *Triticum aestivum* wheat is exclusively grown in **North eastern plain zone**.
218. What is called as King of cereals **Maize**.
219. Critical period for weed competition in maize is **30-45 days**.
220. Cereal which is most suited for saline-alkaline soil is **Barley**.
221. 'Pusa baisakhi' is an important commercial variety of **Mungbean**.
222. Optimum age of seedling for transplanting finger millet is **3-4 weeks**.

223. Richest source of protein among the food grains is **Pulses.**
224. Optimum depth of sowing of Bengal gram **5-8 cm.**
225. Spraying of which of the following conserve moisture in chickpea **$FeSO_4$.**
226. Highest area of sorghum is in **Rajasthan.**
227. Which growth stage of Pigeon pea need bright sunshine for higher yield **Pod development.**
228. Suitable herbicide for pea cultivation is **Isoproturon.**
229. Most nutritious pulse crop is **Lentil.**
230. Pulse crop grown for grain, fodder and green manure is **Cowpea.**
231. Ground nut is sensitive to **Poor drainage.**
232. Age of seedling at transplanting in SRI is. **14**
233. MTU-1001 is a variety of **Rice.**
234. Father of hybrid Rice is **Yuan long ping.**
235. High-yielding varieties of wheat were primarily developed by Indian scientist by crossing- breeding traditional varieties with **Mexican varieties.**
236. How much seed per hectare is needed for Berseem ? **25 - 30 kg**
237. What is the hybrid of Pusa Giant Napier. **Napier × Bajra**
238. Ammonium SO_4 fertilizer is having **21% N and 24% S.**
239. Solar constant **1.94 cal/cm^2/min.**
240. Rainy day is called when **>2.5 mm/day.**
241. NBAIM **Mau.**
242. Family of Sesamum **Pedaliacae.**
243. Certified seeds coloured of tag is **Azure blue.**
244. Biuret % in urea should be **< 1.5%.**
245. Annual RF of world is **1000 mm.**
246. *Lantana Camera* belongs to the Family **Verbanacea.**
247. The herbicide Pendimethalin belongs to the group **Dinitroaniline.**
248. Soil Catena concept given by **Milno.**
249. B.D of earth **2.6-3 g/cc.**
250. Arrangement of Rocks and minerals according to susceptibility to weathering **Weathering Sequence.**

251. If soil formed by deposition with the help of "ICE" is **Morrain.**
252. E-horizon is light in colour because of **Elluviation.**
253. % Pore volume **(1- BD/PD) x 100.**
254. Colloids having size of **< 0.02 mm.**
255. Which clay is tetramorphic **Montmorillonite.**
256. Highly leached and matured soil is **Ultisols.**
257. Cao is a **Quick lime/ Burnt lime.**
258. Sodium Adsorption ratio Equal to **Na/ √ Ca + Mg.**
259. Solan chalk soils are dominated by **Cl^- and SO_4 of Ca, Mg, Na.**
260. Essential commodity Act passed in the year **1955.**
261. Organic Carbon estimation done by **Walkey and Blakley method.**
262. Gypsum Requirement of a soil can be determined by **Schoonovers method.**
263. The word research is derived from **Latin.**
264. NARS –S stands for **System.**
265. d.f. for error in RBD-ANOVA **(r-1) (t-1).**
266. Scientific name of Lentil ***Lens culinaris*.**
267. Bhima and Nari -6 are variety of **Safflower.**
268. Family of Isabgol **Irradiacea.**
269. Depth of sowing will be more in **Marconi wheat.**
270. A short duration crop in between 2 main crop is called **Catch crop.**
271. In Zero tillage the surface soil have **High B.D.**
272. Wind direction is measured by **Anemometer.**
273. SSP contain sulphur **16%.**
274. Inverse yield law proposed by **Wilcox (1929).**
275. What will be CPE Value if 6 cm irrigation is applied at 1.2 IW/CPE ratio **50 mm.**

OOO

CHAPTER 2

Genetics and Plant Breeding

1. First artificial hybrid was produced by **T. Fairchild** by making a cross between **S. William and Carnation Pint.**
2. Cross between Sweet william and Carnation is also called as **Fair Childs mule.**
3. Process of bringing wild species under human management is referred as **Domestication.**
4. Progeny test was given for the 1st time by **Louis D. Vilmorin** while working on sugarbeets.
5. Progeny test is also called as **Vilmorins isolation principle.**
6. Neurotoxin in seeds of Khesari is **BOAA.**
7. First inter-specific hybrid of cotton was **Varalaxmi.**
8. First inter-varietal hybrid of cotton was **H_4**.
9. The selection that leads to speciation is **Disruptive selection.**
10. In cross pollinated species Directional selection leads to **Heterozygous Balance.**
11. In self pollinated species Directional selection leads to **Homozygous balance.**
12. The selection which favors the intermediate phenotype and acts against the extremes is called as **Stabilizing selection.**
13. The hypothesis which explains that allopolyploids evolved by inter-specific hybridization followed by chromosome doubling is **Winge Hypothesis.**
14. Sum total of hereditary material i.e. all the allele of various genes in a crop species and its wild relatives is called as **Germplasm.**

15. Modern cultivars of recent past are called as **Obsolete cultivars.**
16. Gene pool concept was given by **Harlan and Devet.**
17. Gradual loss of variability from cultivated species or wild forms is called as **Genetic erosion.**
18. IRRI, Philippines established in 1960 by the joint efforts of **Ford feller foundation.**
19. In India germplasm collections are maintained by **NBPGR.**
20. Who is known as father of plant genetic resources **N.I. Vavilov.**
21. Who is popularly known as Indian Vavilov **Harbhajan Singh.**
22. The law which explains that characters found in one species also occur in other related species is **Law of Homologous series of Variation.**
23. Within large centers of diversity, small areas that exhibit greater diversity than the centre as a whole is termed as **Micro centers.**
24. Conservation of germplasm in its natural habitat protecting it from human interference is called as ***In-situ* Conservation.**
25. Biosphere reserve is an example for ***In-situ* Conservation** conservation.
26. Conservation of germplasm away from its natural habitat is called as ***Ex-situ* Conservation.**
27. The places with varied climate conditions which leads to different variations among the crop species are called as **Centers of Diversity.**
28. The seeds which are desiccation tolerant and have very long period of storage are called as **Orthodox Seeds.**
29. The seeds which loss their viability if the moisture reduces less than 12-30% is called as **Recalcitrant seeds.**
30. The seed collections that are meant to use actively in breeding programmes is **Working collections.**
31. A minimum set of accessions that together represent genetic diversity is called as **Core collections.**
32. When an introduced material is set to selection or hybridization it is called as **Secondary Introduction.**
33. Isolation of seed materials from spread of diseases, weeds and insect pests is called as **Quarantine.**
34. NBPGR has its head quarters located at **New Delhi.**

35. NBPGR sub-station representing mixed climate is located at **Akola.**
36. NBPGR sub-station representing arid climate is located at **Jodhpur.**
37. NBPGR sub-station representing tropical climate is located at **Kanya Kumari.**
38. NBPGR sub-station representing temperate climate is located at **Shimla.**
39. Introductions of medicinal plants into India will be routed through **BSI.**
40. Adaptation of a variety, line or population to a new environment is called as **Acclamatization.**
41. The phenomenon where seed like structures develop without fertilization is called as **Apomixis.**
42. The modified flowers that develop into plants directly without seed formation is **Bulbils.**
43. If embryos formed from the haploid cells it is called as **Non-recurrent apomixis.**
44. If embryos develop from vegetative cells of ovule like *nucellus*, integument, chalaza without production of embryo sac, it is called as **Adventives embryony.**
45. The cell division followed during the formation of micro and megaspores are **Meiosis.**
46. When the pollen of one flower falls on the stigma of other flowers of the same plant it is called as **Getinogamy.**
47. Failure of pollen of a flower to fertilize same flower is called as **Self-incompatibility.**
48. The incompatibility reaction of pollen is controlled by the genotype of plant which it is produced is called as **Sporophytic self-incompatibility.**
49. The gametophytic system of self incompatibility was 1st described by **East & Mangelsdorf in Nicotiana sandeare.**
50. The progeny obtained from a cross between two facultative apomicts is **Vybrid.**
51. Dominant genes for male sterility are observed in **Safflower.**
52. Production of non-functional pollen grains is **Male sterility.**
53. Transgenic genetic male sterility is used commercially for hybrid seed production in **Crucifers.**

54. TGMS and PGMS are comm... in HSP of which crop for the 1st time **Rice.**
55. Source of male sterile cytoplasm in rice is **Wild Abortive.**
56. Source of male sterile cytoplasm in wheat is ***Ageliops caudata*.**
57. Source of male sterile cytoplasm in maize is **Texas cytoplasm.**
58. Source of male sterile cytoplasm in tobacco is ***N. microcephalis*.**
59. Source of male sterile cytoplasm in bajra is **Tifton21A.**
60. Source of male sterile cytoplasm in jowar is **Combind kaffir.**
61. Source of male sterile cytoplasm in sunflower is ***Helianthus petolaris*.**
62. Source of male sterile cytoplasm in chilli is ***Capsicum annum*.**
63. Source of male sterile cytoplasm in cotton is ***G. harkenssi*.**
64. Source of male sterile cytoplasm in cauliflower is **Ougura cytoplasm.**
65. The male sterility used, where seed is not an economical product/ value is **CMS.**
66. CGMS system is also called as **Nucleoplasmic male sterility.**
67. Most commonly used potential chemical mutagen used for induction of male sterility ***Ethedium bromide*.**
68. Male sterility is transferred by **Backcross method.**
69. The characters which expresses at a particular environment is called as **Threshold character.**
70. The main objectives of hybridization are **Combination breeding, Transgresive breeding & Developing new F_1's.**
71. Removal of stamens or anthers or killing of pollen to prevent selfing is known as **Emasculation.**
72. Breeding for yields generally aims in recovery of **Transgressive segregants.**
73. The law which states that gene and genotypic frequencies will remain constant from generation to generation if there is no selection, mutation, migration or random drift is **H-W law.**
74. Migration is comparable to which system of mating **Genotypic disassortative mating.**
75. Property of an individual to produce progeny is called as **Prepotency.**

76. Hidden or unbound variability which is not available for selection to act upon is **Potential variability.**
77. The three states of variability i.e. free, homozygotic and heterozygotic potential variability are at equilibrium in which system of mating **Random mating.**
78. Reduction or loss of vigor and fertility due to inbreeding is called as **Inbreeding depression.**
79. Sum total of harmful, unfavorable genes constitute the _____ of the species. **Genetic load.**
80. Superiority of F_1 over any of its parents in any character is called as **Heterosis.**
81. Heterosis for the first time coined by **Shull, 1914.**
82. Superiority of F_1 over better parent in any character is called as **Heterobeletosis.**
83. Estimate of Heterosis, which is of commercial or practical value is **Economical Heterosis.**
84. Increased vigor and size of the inter-specific hybrids is called as **Luxuriance.**
85. Luxuriance differs from Heterosis in **Reproductive ability.**
86. Linkage of dominant genes hypothesis was given by **Jones.**
87. Influence of one locus on expression of other is **Epistasis.**
88. Permanent fixation of Heterosis can be done by **Apomixes, Vegetative reproduction and Balanced lethal system.**
89. Purification of existing pureline varieties is done by **Mass selection breeding method.**
90. All available varieties in self pollinated crops are **Purelines.**
91. Progeny of a single self pollinated crop is **Pureline.**
92. Differential rates of reproduction and survival of an organism is called as **Selection.**
93. Rapid advancement of generation of cross is the objective of **Single seed descent method.**
94. Natural selection has more role in which breeding method **Bulk breeding method.**
95. A cross between F_1 to any one of its parents is called as **Backcross method.**

96. The variety developed by back cross method is a **Near isogenic line.**
97. Superiority of F_2 or later generations to the better parent is called as **Transgressive segregation.**
98. The parent with desirable character which is to be transferred to other parent in cross is **Donar parent.**
99. Mixture of purelines with a difference in single R genes is called as **Multilines.**
100. Selection of superior individuals based on the phenotype is **Mass selection.**
101. A tester with broad genetic base i.e. O.P.V or synthetic is used for evaluating lines for **Recurrent selection for GCA.**
102. A tester with narrow genetic base i.e. O.P.V or synthetic is used for evaluating lines for **Recurrent selection for SCA.**
103. Reciprocal recurrent selection method is 1st developed by **Comstock, Robinson and Harvey.**
104. The concept of gamete selection to increase the frequency of superior gametes was given by **Stadler.**
105. Variation observed in *in-vitro* cultured plants is called as **Somaclonal variation.**
106. Use of synthetics in maize was suggested by ____ for the 1st time. **Hayes & Garber.**
107. The ability of inbred to transmit desirable performance to hybrid progenies is called as **Combining ability.**
108. Average performance of an inbred in series of crosses with other inbreds is called **General combining ability.**
109. Excessive performance of a cross over and above the expected performance of GCA of parents is called as **Specific combining ability.**
110. A variety produced by crossing in all possible combinations of a no. of lines that combine well with each other is **Synthetics.**
111. A variety developed by mixing seeds of several phenotypically outstanding lines and encouraging open pollination is **Composites.**
112. Group of plants produced from a single plant through asexual reproduction are called as **Clone.**
113. Clones are said to be immortal, but deteriorate due to **Viral infections and Mutations.**

114. Inability of freshly harvested seeds to germinate for a period of time is known as **Dormancy**.
115. In India the seeds act was enacted in the year _____ by the parliament. **1966.**
116. In India the seeds act came into effect in the year **1969.**
117. The next generation of Breeder seed is known as **Foundation seed.**
118. Registered seed is produced from **Foundation seed.**
119. Certified seed can be produced from **Registered seed or Foundation seed.**
120. The tag colour or label of a Breeder seed is **Yellow.**
121. The tag colour or label of a Foundation seed is **White.**
122. The tag colour or label of a Registered seed is **Purple.**
123. The tag colour or label of a Certified seed is **Blue.**
124. Minimum germination percentage required for certified seeds of Rice is **80%.**
125. _____% of moisture is ideal for certified seeds of Rice **13%.**
126. The newly proposed seed act of India was introduced in **2004.**
127. 'GM' seeds means **Genetically Modified seeds.**
128. The GM crop which was recently at the centre of a controversy in India **Bt Brinjal.**
129. What does 'GEAC' means **Genetic Engineering Approval Committee.**
130. In seed technology, 'PLS' means **Pure Live Seed.**
131. An agro-machinery used for sowing operations **Seed drill.**
132. _____ is the instrument used to measure moisture percentage of seeds **moisture Meter.**
133. _____ is the optimum moisture percentage for storage of seeds of Amaranthus **8%.**
134. _____ is the minimum germination percentage required for certified seeds of Amaranthus **70%.**
135. _____ no. of seeds are normally used in Rag doll method of testing germination **100.**
136. _____ is the capability of seeds to show living properties **Viability.**
137. A chemical used to test the viability of seeds **2,3,5 triphenyl tetrazolium chloride.**
138. 'ICIA' means _____ **International Crop Improvement Association.**

139. National Seeds Corporation(NSC) functions as producer of---class of Seeds **Foundation.**
140. Breeder's seed is otherwise known as _____ **Nucleus Seed.**
141. Foundation seed is otherwise known as _____ **Mother Seed.**
142. Physical purity×Germination percentage/100=? **PLS-Pure Live Seed.**
143. Isolation distance to be followed while cultivating Rice for cerified seeds is _____ **3m.**
144. Genetic purity of Breeder seed is _____ **100%.**
145. Minimum genetic purity required for Foundation seeds is **99%.**
146. Minimum genetic purity required for Certified seeds is _____ **98%.**
147. What's SH medium **Schenk – Hildebrandt medium.**
148. 'Meristem culture' is a method of_____ **Tissue culture.**
149. 'Meristem culture' technique was developed by _____ **George Morel.**
150. _____ can be utilized to produce virus free plant. **Meristem culture.**
151. A trangenic plant containing genes from both potato and tomato plants? **Pomato.**
152. Variation seen in plants produced by plant tissue culture-**Somoclonal variation.**
153. Plants that have been genetically produced using recombinant DNA techniques is called **Transgenic plants.**
154. 'GMO' means? **Genetically Modified Organism.**
155. _____ is useful in Cryopreservation **Liquid Nitrogen.**
156. Sphagnum moss is used as a _____ in Nursery **Growing media.**
157. The Central Seed Testing laboratory is located in **Varanasi & Nagpur.**
158. In India seed certification is **Voluntary.**
159. In India seed labeling is **Compulsory.**
160. The prescribed minimum germination percentage for paddy seed is **80%.**
161. The colour of the tag prescribed for foundation seed **White.**
162. The minimum germination percentage prescribed as for cucumber seed is **60%.**
163. The percent of pure seed prescribed as per minimum seed certification standards (1971) for truthfylly labeled paddy seec **97%.**

164. The isolation distance prescribed for the production of Hybrid Paddy seeds **Foundation Class-200 metres, Certified class 100 metres.**
165. HYVP was state in **1966.**
166. Act 1966 was passed on **29-12-1966.**
167. Identical plants developed from a single mother plant vegetatively are known as **Clones.**
168. _____ is the embryonic root of a plant **Radicle.**
169. The portion of the plant embryo giving rise to the first true leaves is known as **Plumule.**
170. Plants produce seeds that germinate before they detach from the parent are **Viviparous plants.**
171. Scarification is a method to promote **Germination.**
172. _____ is the process of subjecting seeds to both cold and moist conditions to promote germination **Stratification.**
173. _____ is the basis of plant tissue culture **Totipotency.**
174. The concept of Totipotency was originally presented by **Gottlieb Haberlandt.**
175. _____ is the most commonly used medium in plant tissue culture experiments **Murashige & Skoog medium.**
176. Plant part used for micro-propagation **Explant.**
177. In plant tissue culture, 'HEPA' means **High Efficiency Particulate Air.**
178. Plant tissue culture techniques are performed in **Laminar Air Flow Chamber.**
179. Plant hormones that regulate plant growth, particularly by stimulating cell Elongation **Auxin.**
180. Naturally occurring Auxin is **IAA.**
181. _____ are a class of plant hormones that promote cell division in plant roots and Shoots **Cytokinin.**
182. _____ is a plant hormone used to promote stem elongation& seed Germination **Gibberellin.**
183. _____ is known as the 'plant stress hormone' causing inhibition to plant growth **ABA.**
184. _____ is a plant hormone in gaseous form **Ethylene.**
185. Who developed pH scale **S.P.L Sorensen.**

186. _____ is a gelling agent in culture medium **Agar.**
187. The National Seeds Corporation (NSC) was constituted in the year **1963.**
188. _____ is the fundamental unit of crop production **Seed.**
189. What is meant by HYV **High Yielding Varieties.**
190. _____ is the progeny of breeder seed **Foundation Seed.**
191. Certified seed is produced from **Foundation Seed/Registered Seed.**
192. In seed technology, 'TLS' means **Truthfully Labelled Seeds.**
193. 'GATT' means **General Agreement on Tariffs and Trade.**
194. 'GATT' agreement was came into force in **1948.**
195. 'WTO' means **World Trade Organisation.**
196. World Trade Organisation was formed in the year **1995.**
197. Expand the term "TRIPS" **Trade Related Intellectual Property Rights.**
198. In seed technology, 'PBR' means **Plant Breeders Right.**
199. NBPGR (National Bureau of Plant Genetic Resources) is located in **New Delhi.**
200. Expand UNEP **United Nations Environment Programme.**
201. _____ is a mature ovule containing an embryo **Seed.**
202. In a seed the food storage tissue is known as _____ **Endosperm/ Cotyledon/ Perisperm.**
203. World's first seed testing laboratory was established in _____ in 1869 **Germany.**
204. International Seed Testing Association (ISTA) was established in **1924.**
205. _____ is a method of testing seed germination **Rag Doll Method/ Paper Method.**
206. In Rag doll method of testing seed germination, flannel clothes used is of size **35cm × 27.5cm.**
207. An agro-machinery used for sowing operation **Seed Drill.**
208. _____ is the instrument used to measure moisture percentage of seeds **Moisture Meter.**
209. Generally asexually propagated crops are perennial in nature, but **Cassava** is a annual clone crop.
210. Ideotype concept was developed by **Donald.**

211. A biological model expected to perform in a predictable manner within a defined environment is called as **Ideotype.**
212. The Ideotype which performs best in commercial crop densities is **Crop Ideotype.**
213. The Ideotype for irrigated cotton was given by **Singh.**
214. Term mutation is given by **Hugo de veris.**
215. Example for a non-ionizing physical mutagen is **U.V rays.**
216. Only recessive traits can be studied with **Lethal mutations,** which are based on effect of survival.
217. Majority of mutations are **Harmful** in nature.
218. Plants that owe their resistance by avoidance are called as **Water savers.**
219. Resistance against chilling can be conferred by accumulation of **ABA** in plant tissues.
220. Formation of ice crystals inter and intra cellular level is due to **Freezing** stress.
221. Capacity of a pathogen to incite a disease is called as **Virulence.**
222. When the rate of reproduction of a pathogen is equal to 0 i.e. $r = 0$ the reaction is **Immunity.**
223. Gene for gene hypothesis was given by **Flor.**
224. When pathogen genes are matched with all resistance genes in the host, it produces **Susceptibility.**
225. Genetics of disease resistance is given by **Biffen.**
226. Gene for gene hypothesis was given by flor while working on **Linseed rust.**
227. Boom and bust cycle of pest outbreak was given by **Priestly.**
228. Mechanisms of insect resistance are given by **Painter.**
229. High silica content in rice leaves and stem confers resistance against **Yellow stem borer.**
230. DIMBOA in leaves in maize confers resistance against **European corn borer.**
231. Detrimental effect of host plant on the development of insect is called as **Antibiosis.**
232. Effect of pollen grains on the maternal tissues is **Metaxenia.**
233. Process in which pollination and fertilization takes place but developing embryos abort after sometime is **Stenospermocarpy.**

234. An abnormal condition produced by an organism is called as **Disease**.
235. Ability of a host to avoid or minimize loss in productivity instead of its infection by a pathogen is **Tolerance**.
236. Escape of a host plant from damage is called as **Avoidance**.
237. High benzyl alcohol in wheat confers resistance against **Green bugs**.
238. Frego bract in cotton confers resistance against **Boll weevil**.
239. In Initial Evaluation Trail (IET's) the trails are conducted in **10-12** locations.
240. In Uniform Regional Trails (URT's) the trails are conducted at **25-30 locations**.
241. Adoptive Research Trails are conducted in **0.5 acre** area.
242. Y- Radiated papaya variety commercialized is **Pusa nanha**.
243. In random mating populations the inbreeding coefficient 'F' is **0**.
244. The process of Quarantine takes at least **3 weeks** time.
245. Hybrids from cross between two or more inbreds or purelines are called as **Conventional hybrids**.
246. XYZ system of Hybrid seed production is developed by **Drisscoll**.
247. Segregating materials grown in different locations to facilitate selection for one trait in one location and for another trait in other is called as **Shuttle breeding**.
248. Presence of physical barrier between anthers and stigma which promotes cross pollination is **Herkogamy**.
249. Loss of a genotype due to its failure to reach an optimum phenotype due to GxE interacation is called as **Genetic slippage**.
250. Mass selection is based on **Maternal** parent only.
251. Proportion of variability expressed as phenotype is **Free variability**.
252. Heterozygotic potential variability released as free variability by **Segregation**.
253. Hybrid necrosis in wheat is due to **Ne-1, Ne-2** genes located on $5B^L$ & $2B^S$ chromosomes.
254. Clones have identical genotypes but not in **Phenotype**.
255. The host effected by pathogen but does not show any symptoms are called as **Resistance**.
256. Instrument used to measure tissue water potential is **Psycrometer**.

257. Defect correction can be done by **Backcross method.**
258. **Completion Ideotype** Ideotype is favored in segregating generation of crosses.
259. Selection for one trait would reduce the yield is referred as **Yield penalty**.
260. Rapeseed varieties with low erucic acid are called as **Single zero varieties.**
261. Rapeseed varieties with low erucic acid and low glucosinolates are called as **Double zero varieties.**
262. Rapeseed varieties with low erucic acid, low glucosinolates and low fibre are called as **Triple zero varieties.**
263. **Mixogram** test in wheat indicates good baking quality.
264. Trait not observed in wild relative but observed in segregates' from crosses of crop species with wild relative is **Latent trait**.
265. Strong gluten strength is required for **Bread making**.
266. Source is limited in **Pulses.**
267. Sink is limited in **Cereals.**
268. The Ideotype that performs well in genetically heterogenous population is **Competition ideotype**.
269. Selection in one direction for one generation and in opposite direction in another generation is **Cyclic selection**.
270. Diallel selective mating design is given by **Jenson.**
271. Sib pollination with stigma and pollen of two different individuals that are vegetatively derived from same mother is called as **Adelphogamy**.
272. Techniques used to promote homologous pairing & translocations is **Chromosome manipulation techniques.**
273. A group of species with common evolutionary ancestry is called as **Clade.**
274. The difference in behavior or activity between the alleles contributing by 2 parents in an individual is **Genetic imprinting**.
275. Homologous genes within same organism and encode proteins that are related and have identical functions are called as **Orthologues**.
276. Original plant from which a clone is taken is called as **Orlet.**
277. Tendency to eliminate undesirable genotypes from the population is called as **Selection pressure.**

278. Complete failure of vertical resistance leads to a disease epidemic called as **Vertifolia effect**.
279. Storage protein in maize is **Zein**.
280. Drosophila of plant kingdom is **Arabidopsis**.
281. Strategic use of major genes by deploying them over space and time to avoid phenomenon of boom and bust is referred as **Gene deployment**.
282. Mutant allele that functions more effectively than a normal allele is **Hypermorph**.
283. Competition between gametes carrying mutant and normal alleles is called as **Haplontic selection**.
284. Severe outbreak of a disease from a low level of infection is called as **Epidemic**.
285. Occurrence of a disease at irregular intervals /locations and at low frequency is called as **Sporadic disease**.
286. A disease more or less constantly present from year to year in moderate severe form in a locality is **Endemic disease**.
287. The hypothesis which explains that horizontal resistance is the cumulative effect of those vertical resistance genes which succumbed to corresponding virulence genes of pathogens is **Ghost gene hypothesis**.
288. Isolation of hybrids and developing haploids by anthers/ovary culture and then double haploids is **Reverse breeding**.
289. An undesirable consequence due to close linkage between a desirable and an undesirable gene is **Linkage drag**.
290. Lines identical in genotype except for one gene are called as **Near isogenic lines**.
291. Scheme according to which individuals are mated to produce a sexual progeny is **Mating system**.
292. Dehiscence of anthers of flowers before stigma is receptive is called as **Protandry**.
293. Vertifolia is a variety of **Potato**.
294. Use of genetic factors to make male gamete non-functional is called as **Genetic emasculation**.
295. The term genetic resources was coined by **Frankel**.
296. Mendel began his famous experiments on **Pea in 1857**.

297. Mendel Paper entitled "Experiments in Plant Hybridization" was presented **in German Language in two papers in 1865 (Feb 8 and March 8).**
298. The term "Genetics" was coined by **Bateson (1905).**
299. The term gene, genotype and phenotype were used by **Johannsen in 1903.**
300. Germplasm theory was proposed by **Weismann.**
301. Crossing over term was used first time by **Morgan and cattell.**
302. Operon model of gene regulation was discovered by **Jacob and Monad 1961.**
303. The term chromosome was coined by **Waldeyer in 1888.**
304. One gene one enzmes hypothesis was proposed by **Beadle and Tatum in 1941.**
305. First time gene was divided into **cistron, muton and recon by Benzer** (1955).
306. Sex chromosomes are also known as **allosomes.**
307. Crossing over occurs in **Pachytene stage.**
308. The tendency of two or more genes to remain together in the same chromsome during inheritance is referred to as **Linkage.**
309. Chiasma type theory of crossing over was proposed by **Janssens.**
310. Genic balance system of sex determination was proposed by **Bridges in 1922.**
311. Monogenic sex determination is found in - **Asparagus, Papaya, maize, spinach etc.**
312. The concept of purelines theory was developed by **-Johannsen (1903), Danish biologist.**
313. Breeding method for self -pollinated crops are - **Pure line selection, Mass Selection, Progeny selection, Bulk Method, Pedigree method, single seed descent method, Back Cross Method.**
314. Bulk method was first used by **Nelson Ehle in 1908.**
315. The modification of bulk method is **Single seed descent.**
316. In pure line Theory, Johannsen was working on **Princess variety of Rajma.**
317. Mass Pedigree Method was proposed by **Harrington in 1937.**
318. The oldest method of selection is **Mass Selection.**
319. Progeny Test is also known as **Vilmorin isolation principle.**

320. Single seed Descent method applied for the first time in oat in 1965 by **Graphius.**
321. The term recurrent selection was coined by **Hull in 1945.**
322. The use of synthetic varieties for commercial cultivation was first suggested in maize by **Hayes and Garber, 1919.**
323. Synthetic varieties are maintained by **Open Pollination.**
324. A synthetic variety can be developed from **Inbreeds, clones and open pollinated varieties.**
325. The general or basic concept in the development of synthetic varieties is **Exploitation of hybrid vigour.**
326. The concept of disruptive mating and selection was developed by **Mather and Thody (1958).**
327. The heterosis is also known as **Hybrid vigour.**
328. Dominance theory was proposed by **Devenport in 1908.**
329. The cross between inbred line and open pollinated variety was done in **Top cross.**
330. The resistance of host to the particular race of a pathogen is known as **Vertical resistance.**
331. Vertical resistance is governed by **One or few genes.**
332. The resistance of host to the all race of a pathogen is known as **Horizontal resistance.**
333. The disease resistance is governed by **Several genes.**
334. Who developed the concept of vertical and horizontal resistance **J. E. Vender Plank.**
335. Father of genetic engineering is **Pual Berg.**
336. Polymerase chain reaction (PCR) was developed by **Dr. Karl Mullis (1980).**
337. Golden rice was developed by **Ingo protrykus and peter bayer.**
338. The plant part , which is used for regeneration is called **Explant.**
339. A mass of regenerated cells i culture medium is called **Callus.**
340. The suspension of free cells of callus in a liquid medium is known as **Suspension culture.**
341. The regeneration capacity or ability of a plant cell to develop into whole plant is known as **Totipotency.**
342. What is the embryogenesis - **It is the process of formation of somatic embryos from callus.**

343. The process of dirrerentiation of shoot & root from the somatic embyos is called **Organogenesis.**
344. Regeneration of whole plant from anther is called as **Anther culture.**
345. Haploid are developed by **Anther culture.**
346. The Blue biotechnology is a term that has been used to describe - **The marine and aquatic application.**
347. The biotechnology applied to agriculture process is known as **Industrial Biotechnology.**
348. The first genetically modified food product was a **Tomato, which was transformed to delay its ripening.**
349. m-RNA is primarily used for **Transcription of DNA.**
350. RFLP stands for - **Restricting fragment length polymorphism.**
351. Plant tissue culture was begun by **G. Haberlandt (geman botanist) in 1808.**
352. Research in the area of tissue culture was first stared in India at **Department of Botany, DU in 1960s.**
353. The project on Indian mustard oil with higher β-carotene is initiated by **TERI.**
354. A genotype developed by the process of genetic engineering is called **Transgenic.**
355. **What is gene cloning** - A technique of genetic engineering, which is used to make several identical copies of a gene is called gene cloning.
356. Which soil borne bacterium is used for development of transgenic plants ***Agrobacterium tumifaciens***.
357. In potato, the protein quality has been improved by transferring serum albumin gene from **Human.**
358. In tobacco, cold resistance power has been achieved by transferring gene from ***Arabidopsis thaliana***.
359. For every generation of selfing heterozygosity is reduced at a rate of **50%.**
360. Triticale is a cross between **Wheat and rye.**
361. Source of rice dwarfing gene is from **Dee geo woo gen** variety.
362. What is possible number of straight crosses along with reciprocals with 10 inbred lines is **90.**

363. Embryonic root inside seed is called as **Radicle.**
364. Physical basis of life by Huxely is **Protoplasm.**
365. Centrioles are found in **Animals** only.
366. Expression of transgenes in plants is effected by **Gene silencing, Chromosome arrangement, Pleiotrophy**.
367. Removal of RNA primer during DNA replication in prokaryotes is done by **DNA pol-1**.
368. Natural genetic engineer is **Agrobacterium.**
369. Genes subject to dosage compensation are concerned with **Embryo development.**
370. Genetic transformation in animal cells is **Transfection.**
371. Transformation experiments were carried by **Griffith.**
372. Discovery of DNA as genetic material was done by **Avery, MaCleod and McCarty**.
373. All viruses are **Obligate** parasites.
374. Single stranded RNA as genetic material is seen in **Polio, influenza.**
375. De oxy ribose sugar lacks oxygen atom at **2^1 carbon position.**
376. Nitrogen base + Pentose sugar are together called as **Nucleosides.**
377. Nitrogen base + Pentose sugar along with phosphate group is called as **Nucleotides**.
378. The two strands of DNA helix are held together by **Hydrophobic forces.**
379. X-ray crystallography of purified DNA is revealed by **Wilkins, Franklin.**
380. **Complementary base pairing** yields error free replication of DNA.
381. DNA-RNA heteroduplexes and RNA double helixes occur in-vivo in **A-Form** DNA.
382. Left hand coiling is seen in **Z-form** DNA.
383. Semi-conservative replication was postulated by **Watson and Crick.**
384. DNA molecule that undergoes replication is **Template.**
385. Autoradiograph of replication in E. coli was shown by **J. Cairns.**
386. Rolling circle replication of phages is discovered by **Dresseler**.
387. The lagging strand replication is discontinuous and produces polynucleotide fragments called as **Okazaki fragments.**

388. DNA polymerase needs **3' OH** site to initiate replication.
389. **DNA polymerase-1** Enzyme is called as korenberg enzyme.
390. Repair by breaking thymine dimmers by Phy gene product in the presence of blue light is called as **Photoreactivation.**
391. The technique used to detect the presence of given sequence of DNA/RNA in sample of DNA's is **DOT Blot technique.**
392. Characterizing of large regions of chromosomes is by **Chromosome walking.**
393. Father of biochemical genetics **Garrod.**
394. One gene one polypeptide hypothesis is by **Ingram.**
395. Deciphering of genetic code is done by **Nirenberg, Ochoa, Khorana.**
396. In-vitro synthesis of DNA is by **Khorana.**
397. In-vitro synthesis of RNA is by **Ochoa.**
398. Amino acids with single codon each are **Trypthophan, Methionine.**
399. UV rays have its highest mutagenic effects at **254 Å** wavelength.
400. Cross-linking of DNA is due to **Alkalyting agents.**
401. Substitution of one purine by another pyramidine is called as **Transversion.**
402. Substitution of one purine by another purine or pyramidine by other pyramidine is called as **Transition.**
403. Paramutations are due to **Heterochromatization or Metylation** of DNA.
404. **Duplications** Structural aberrations are the basis for evolution.
405. The inversion where centromere is involved in the inverted segment is **Pericentric inversion.**
406. Group of symptoms leads to **Syndrome.**
407. Acentric and dicentric fragments are the result of a single crossover in **Paracentric inversions.**
408. Functional gametes are formed only due to **Alternate** segregation.
409. When non-homologous centromeres move to one pole it is **Adjacent-1** disjunction.
410. Permanent heterozygosity as in oenothera can be maintained **Regular alternate disjunction, Balanced lethal system.**
411. How many linkage groups are possible in any organism **n haploid chromosome number.**

412. The extra chromosome in a primary trisomic is an **Normal chromosome.**
413. The extra chromosome in a secondary trisomic is an **Isochromosome.**
414. The extra chromosome in a teritory trisomic is an **Translocated chromosome.**
415. The possible number of secondary trisomics in rice is **2n.**
416. The possible number of primary and secondary trisomics is **n.**
417. Phenotypically males but are females is the characteristic feature of **Klinfelter's syndrome.**
418. Trisomy of chromosome 13 in humans leads to **Patau's syndrome.**
419. Trisomy of chromosome 18 in humans leads to **Edward's syndrome.**
420. Trisomy of chromosome 21 in humans leads to **Down's syndrome.**
421. Pseudo dominance is due to **Deletions.**
422. Monosomy of X chromosomes in humans leads to **Turners syndrome.**
423. **Quinacrin mustard** stain is used to produce Q-Bands.
424. Traffic police of the cells is **Golgi complex.**
425. Significance of nucleus was given by **Hammerling.**
426. Nucleus is absent in **Companion cells of phloem, RBC's.**
427. Only one large chloroplast is seen in the cells of **Chlamydomonas.**
428. Lowest number of chromosomes in plant is seen in ***Haplopappus gracilis*.**
429. Chromosomes are seen as smallest in **Anaphase.**
430. Longest chromosome is seen in **Trillium sp.**
431. Bead like structures seen during prophases is called as **Chromomere.**
432. Arrangement of chromosomes in descending order of size is **Karyotype.**
433. Diagrammatic representation of karyotype is **Ideotype.**
434. Polycentric chromosomes are seen in ***Luzule* sp.**
435. The chromosomes are seen as J shaped during metaphase are **Acrocentric chromosome.**
436. Site of spindle fibre attachment is called as **Centromere.**
437. Histones that are rich in lysine amino acid are **H_1, H_2a, H_2b.**
438. Histones that are rich in arginine amino acid are **H_3, H_4.**
439. Individual integrity of chromosomes are maintained by **Telomeres.**

440. Telomeres have **3[1]TTAGGG5[1]** sequence in humans.
441. Folded fibre model was given by **Dupraw.**
442. Nucleosome solenoid model was given by **Korenberg and Thomas.**
443. Interphase is absent in ***Trillium* sp.**
444. Chromosomes are shortest and thickest in **Metaphase.**
445. Sister chromatids will move to opposite poles during **Anaphase-II** of meiosis.
446. Chaisma terminilization starts in **Diplotene** stage of prophase-I.
447. Effect of one gene on expression of other gene is **Epistasis.**
448. Tendency of 2 or more genes to stay together during inheritance is **Linkage.**
449. Linkage is detected by **Test cross.**
450. Exchange of homologous segments between non-sister chromatids of homologous chromosomes is **Crossing over.**
451. Complete linkage is seen in **Drosophila males and silkworm females.**
452. 1 mapunit or 1 cm is equal to **1000 kb DNA in man.**
453. Occurrence of one crossing over in region suppresses occurrence of another crossingover in its vicinity is **Interference.**
454. The ratio between the observed frequencies of Double crossovers to expected crossovers is **Coincidence.**
455. Griffith's transformation experiment was conducted on ***Diplococcus pneumonia*** organism.
456. Buffering capacity of genotype to environmental fluctuations is called as **Homeostasis.**
457. When RBC's have antigen A on their surface and antibody B on serum the blood group is **A Blood group.**
458. Characters of sex linked inheritance are observed high in **Heterogametic sex.**
459. OMMI stands for **Online Mendelian Inheritance in Man.**
460. The first transgenic plant tobacco was developed resistance against **Kanamycin.**
461. 2:1:1 ratio of testcross progeny is observed in **Dominant Epistasis.**
462. Which technique is used to minimize the expression of gene **Antisence RNA Technology.**

463. DNA sequences that increase the function of promoter is called as **Enhancer.**
464. Different alleles produce a same phenotype is **Isoalleles.**
465. Unstained and dividing cells are studied using **Phase contrast microscope.**
466. Acid phosphatase acts as marker enzyme in **Lysosome.**
467. At start of transcription 2-9 oligo-nucleotides formed is **Abortive transcription.**
468. Process of generation of mature RNA is called as **Splicing.**
469. Highly mutable sites within the gene are **Hotspots.**
470. Gene placed on both physical and genetic map is **Anchor gene.**
471. Appereance of wild phenotype in F_1 from cross between two mutants is **Complementation.**
472. In Haploid organisms, non-dividing cells are blocked in **G2 phase.**
473. In Diploid organisms, non-dividing cells are blocked in **G1 phase.**
474. One allele modify other alleles on same nucleus in such a way that function of other allele remains modified for few generations is called as **Paramutation.**
475. Negative interference is observed in **Viruses.**
476. Partially homologous chromosomes are called as **Homeologous.**
477. Unit of lignin is **Coniferyl alcohol.**
478. Binding of regulatory protein to operator is necessary for transcription to begin in case of **Positive control.**
479. Development of average phenotype irrespective of environmental or genetic variation is **Canalization.**
480. Premature termination of transcription is called as **Attenuation.**
481. Twins showing similar phenotypes for a trait is called as **Concordance.**
482. Gamma ray dose is expressed as **Gray.**
483. Marker protein for mitochondria is **Cytochrome-C.**
484. Gene families are originated by duplications followed by **Mutations.**
485. The Dot blot technique helps in detection of **DNA sequence.**
486. Haploid with an iso-chromosome is called as **Mis-division Haploid.**

487. Deficiency in chromosome 5 in humans leads to **Cri-du-chat syndrome**.

488. The individual with AAXX are **Females**.

489. In sickle cell anemia, B-chain has valine instead of **Glutamine** at position 6.

490. Promoter region can be identified by **foot printing**.

491. In papaya, the genotype M_1m will be phenotypically **Males**.

492. The interval mapping used where information about number, position and effects of QTL's can be known as **Baysian interval mapping**.

493. Metals used in particle bombardment method are **Gold, Tungsten**.

OOO

CHAPTER 3

Agricultural Microbiology

History of Microbiology

1. Father of Microbiology **Louis Pasteur.**
2. Father of Soil Microbiology **Winogradsky.**
3. **Leeuwenhock** invented 1st simple microscope.
4. He described the microbial world he observed as **'animal cules' or 'little animals'** or small animals.
5. **Robert Hook/**Father of cell theory, invented 1st compound microscope.
6. **Ruska, Knolland & Borris** invented Electron microscope.
7. **Abiogenesis** Origin of life from non-living matter also called as spontaneous generation theory.
8. **Biogenesis** origin of life from other living organism.
9. **Franceso Redi:** performed experiments and showed that maggots that develop in putrefying meat are the larval stages of flies and will never develop in putrefying meat if it is protected from flies laying eggs.
10. He was the first to disprove opponent of **Abiogenesis called doctrine of spontaneous generation or Abiogenesis.**
11. **Lazzaro Spallanzani (1729-99)** was the first to provide evidence that microorganisms do not develop spontaneously.
12. **Louis Pasteur:** Founder of 'germ theory of disease'.
13. Proved **Pebrine disease** of silkworm is caused by **protozoa.**
14. Anthrax is the first disease of animal proved to be caused by bacterium.
15. Pasteurization **(62.8°C (145°F) for ½ hour).**

16. Discovered the terms **aerobic & anaerobic** during **butyric acid fermentation** also discovered the existence of life in presence of O_2 and gave the term facultative anaerobic.
17. Sugar Yeast Alchol + Co_2 $\xrightarrow{}$ Sugar Yeast Alchol + Co_2
    $-O_2$ $+O_2$

Robert Koch

1. Identified bacteria causing anthrax is ***Bacillus anthracis***.
2. Developed pure culture techniques (**Streak and pour plate**).
3. Discovered TB caused by ***Mycobacterium tuberculoses***.
4. Cholera caused by ***spirillum*** **bacteria**.
5. Developed **gelatin plate technique** for isolation of bacteria.
6. Used gelatin as solidifying agent which was replaced by **Agar – agar**.
7. **Buchner:** Complex metabolic process of fermentation was brought about by catalyst in the yeast cells which is protein & he named it as '**Zymase**'.
8. **Brefeld:** Worked on **pure culture** of fungi technique.

Joseph lister

1. **Obtain first time pure culture of milk bacteria (*Bacillus lactic*) by using serial dilution method.**
2. **Used carbolic acid (Phenol) for disinfecting surgical dressing.**
3. Developed 'aseptic technique' known as father of **aseptic surgery**.

Winogradsky & Beijerinck

1. Developed 'Enrichment culture technique'.
2. **Enrichment culture is most powerful task for isolation of any type of microbe.**
3. **Paul Ehrlick:** Father of chemotherapy.

Wingradsky

1. Father of soil microbiology.
2. Role of bacteria in nitrification.
3. Phenomenon of antagonism in soil microbe.
4. Used the term chemoautotrophs.

Waksman

1. Gave the term antibiotic and said actinomycetes are prolific producers of antibiotics.
2. Discovery of streptomycin antibiotic.
3. **Berkely:** Potato blight caused by **fungus.**
4. **Alexander flemming:** Fungi produce antibiotics and he discovered penicillin antibiotic.
5. **Hellriegel & Wilfrath:** Mutual beneficial association between roots of legume and rhizobia.
6. **Bouiosingault**: Legume fixes atmospheric nitrogen and increase nitrogen content in soil.
7. **Frank**: Discovered actinomycetes 'Frankia' and gave the term mycorrhiza.
8. **Beijerinck**: Father of microbial ecology and discovered Azotobacter.
9. **Hiltner**: Coined the term ' Rhizosphare'.
10. **Edward Jenner**: Father of immunology developed vaccines.
11. He developed concept of **vaccination** against small pox by vaccinating them with attenuated cow-pox.
12. Growth in bacteria defined as increase in cell number in a given unit time.
13. During bacteria growth, cell increase in number per population by geometric progression.
14. **Generation time (G):** Time required for a cell to divide of a population to double.
    a. *Ex: E. coli* = 20 min,
    b. *Bacillus subtilis* = 25 min,
    c. *M. tuberculosis* = 12 hr
15. **Growth rate (R):** number of generation /division per unit time
    a. R ∞ 1/G.
16. **E.H. Haeckel in 1866** proposed a third kingdom **'Protista'** to include the microorganisms that are typically neither plants nor animals.
17. Bacteria are referred to as **lower protists.**
18. The fungi, algae and protozoa are called **higher protists.**

19. **Prokaryotes** (incipient nucleus) and **eukaryotes** (true nucleus).
20. Bacteria are prokaryotic microorganisms and are called as prokaryotes.
21. Algae, fungi and protozoa are eukaryotic microorganisms and are referred as **eukaryotes** (plant and animal cells are also eukaryotic).
22. The eukaryotic cells are characterized by the presence of **multiplicity of unit membrane systems,** which are structurally and topographically distinct from cytoplasmic membrane.
23. Surface area/ volume ratio is exceedingly high favoring unusually **high rate** of growth and metabolism of bacteria.
24. Trichomes producing organism Ex: *Beggiatoa.*
25. **Palisade** arrangement (*Corynebacterium diphtheria*).
26. Bacterial flagella are hair like **helical** appendages that protrude through the cell wall and are responsible for **swimming motility**.
27. The protein of the filament is known as **flagellin.**
28. **Monotrichous** – a single polar flagellum.
29. **Lophotrichous** – a cluster of polar flagella.
30. **Amphitrichous** – flagella either single or clusters, at both cell poles and
31. **Peritrichous** –surrounded by lateral flagella.
32. Without external flagella some helical bacteria **(spirochetes)** exhibit swimming motility by means of **Endoflagella** (flagella like structures beneath the outer cell envelope).
33. Some bacteria ex **myxobacteria** exhibit **gliding motility** only when they are in contact with solid surface.
34. Bacterial **Chemotaxis** is the movement of bacteria toward or away from the chemical compounds.
35. Swimming of bacteria towards a chemical is known as **positive chemotaxis.**
36. Swimming away a chemical is known as is **negative chemotaxis**.
37. **Phototrophic bacteria** exhibit positive phototaxis toward increasing light intensitie and are repelled by decreasing light intensities.

Pili (Fimbriae)

1. They are **hollow non-helical filamentous** appendages that are **thinner, shorter** and more **numerous** than flagella.

2. **F – pilus** (Sex pilus) serves as the port of entry of genetic material during **bacterial mating.**
3. **Capsule:** Capsule made up of **organic exopolymers** that form an envelope outside the cell wall.
4. In bacteria the **cell wall** is very **rigid** and gives the **shape** to the cell.
5. **Eubacteria** cell wall is made up of **peptidoglycan** murein and insoluble, porous cross linked polymer of enormous strength and rigidity.
6. **Peptidoglycan** is basically a polymer of **N-acetylglucosamine, N-acetylmuramic acid, L- alanine, D-alanine, D-glutamate and a diamino acid.**
7. The peptidoglycan is present only in **prokaryotes.**
8. The cell walls of **archaebacteria** are generally made up of **proteins, glycoproteins or polysaccharides.**
9. **Gram staining** is one of the most important and widely used differential staining introduced by **Christian Gram in 1884**.
10. Bacteria stained by **Gram's staining** method fall into two groups–

    (*i*) Gram positive -which appear deep violet in colour

    (*ii*) Gram negative - which appear red in colour.
11. **Gram positive bacteria** by treating the cells with an enzyme such as **lysozyme,** which selectively dissolves the **cell wall** or by culturing the bacteria in the presence of an **antibiotic such as penicillin.**
12. Sphaeroplast is a protoplast surrounded by the outer membrane of cell wall.
13. In **gram-negative bacteria** only **peptidoglycan** layer can be removed but outer membrane is still intact surrounding the protoplast.
14. In many bacteria, especially **Gram-positive bacteria,** the cytoplasmic membrane appears to be infolded at more than one point. Such infoldings are called **mesosomes.**
15. Mesosomes are thought to be involved in Respiration, **DNA replication, cell division** and **export of exocellular enzymes.**
16. The bacterial nucleus is **not enclosed** in a defined **membranous structure.**
17. The bacterial **cell contain single circular double stranded DNA molecule.**

18. This nuclear material is generally designated as **nucleoid,** or **chromatin body** or **nuclear equivalent.**
19. Ribosomes are **70 S type** consisting of **50 S** and **30 S sub-units.**
20. **Volutin granules** reserve source of phosphate.
21. **poly-β-hydroxybutyrate** (PHB) and **glycogen** both serving as source of carbon and energy.
22. Spores produced within the cell are called **endospores.**
23. Spores produced external to cell are called **exospores.**
24. Endospores are **thick walled, highly refractile bodies** that are produced (one per cell) *by* ***Bacillus, Clostridium, Sporosarcina*** and few other genera.
25. They are generally formed at the end of the active growth or during **stationary phase.**
26. Exospores are formed external to the vegetative cell by budding at one end of the cell in the **methane oxidizing genus** ***Methylosinus***.
27. ***Azotobacter*** sp. produce cysts.
28. Budding bacteria are ***Rhodopseudomonas, Hyphomicrobium***.
29. Pure culture technique for bacteria was developed by **Koch.**
30. Sterols are absent in the cytoplasmic membrane of **Prokaryotes.**
31. Secondary metabolites are produced during **Stationary** phase.
32. The chemical nature of the flagellum is **Protein (Flagellin).**
33. The plasmid upon integration onto the bacterial genome is called **Episome.**
34. Antibiotic Streptomycin was discovered by **Waksman.**
35. Most of the bacteria grow at a pH of **6.8-7 (Neutral).**
36. Most of the bacteria grow at a pH **Acidic Nature (3-4).**
37. Bacteria that require essentially Oxygen in low concentrations for their optimal growth are **Micro Aerophiles.**
38. The process of extraction of minerals from low grade ores is called **Geo Chemical Microbiology.**
39. Primary metabolites are produced during **Lag/ Exponential** phase.
40. Extra chromosomal elements in bacteria are known as **Plasmid.**
41. Goose neck flask was invented by Louis Pasteur to disprove the theory of **Spontaneous generation.**
42. The increase in the cell mass or cell number per unit time is called **Growth rate.**

43. Special defenders against infection as Phagocytes (eating cells) and the process **'Phagocytosis'**.
44. Phagocytosis is discovered by **Elie Metchnikoff.**
45. The biological specifity of diseases was demonstrated by **Robert Koch.**
46. Bacteria that requires additional growth factors for their growth are **Auxotrophs.**
47. An example of aerobic endospore former is ***Bacillus sp.***
48. An example of anerobic endospore former is ***Clostridium.***
49. Example of Exospores former is **methane oxidizing genus *Methylosinus*.**
50. Cysts are formed by the bacterial genus ***Azotobacter.***
51. **Joseph Lister** developed antiseptic method of surgery.
52. Volutin granules in bacteria serves as a reserve source of **phosphate (P).**
53. Endospores are resistant to high temperature & chemicals because of the presence of a chemical, **calcium dipicolinic acid.**
54. The kingdom Protista for including microorganisms was proposed by **Haeckel.**
55. Variability in bacteria due to environmental factors is known as **Adaptation.**
56. Antony Von Leeuwenhoek described microbes as **Animalcules.**
57. A method of sterilization where discontinuous heating is used is called **Tyndalization.**
58. One gene one enzyme hypothesis proposed by **Beedle and Tatum.**
59. Bacteria that cannot grow in presence of oxygen are called as **Obligate Anerobe.**
60. In bacteria, bunch of cells is called **staphylo (coccus/bacilli).**
61. In true photosynthesis, the electron donor is **Water (H_2O).**
62. Antibiotics are produced by bacteria when they are in **Stationary** phase of growth.
63. Bacteria that require essentially small quantities of oxygen are **Micro aerophiles.**
64. Example of Methanogenic bacteria ***Methanosarcina, Methanothrix, Methanobacterium*** **and** ***Methanospirillum.***

65. **Microaerophilic bacteria** Ex. *Azospirillum, Spirosoma, Campylobater.*
66. **Facultative anaerobic bacteria** Ex. *E.coli, Erwinia, Enterbactor.*
67. **Acidophiles** - pH < 5.
68. **Alcaliphiles** - pH 7 to 12.
69. **Psychrophiles** = optimum temperature **15°C range = 0-5°C.**
70. **Mesophiles** = optimum temperature **37°C range = 25-40°C.**
    (*a*) Therefore are bacteria pathogenic to human & animals are mesophilic in nature.
    (*b*) Mesophiles & Psychrophiles killed at pasteurization temperature.
71. **Thermophiles** - optimum temperature **60-65°C range = 45-80°C.**
72. Extreme thermophiles Ex. ***Archaebacteria (Pyrodictium)*** are capable of growing even at 100 - 110°C.
73. Most of the bacteria grow optimum pH 6.5 to 7.5.
74. **Osmophiles** (high sugar concentration) ex: Yeast, mold.
75. **Halophiles** (high salt concentration) ex: Halobacterium (Extreme halophiles).
76. The structure responsible for respiration in bacteria is **Mesosomes.**
77. Kingdom prokaryote includes **Bacteria, Cynobacteria.**
78. *E. coil* is grown in a medium containing glucose & lactose, it shows **Diauxic** growth.
79. Biogas end product consists of **Methane (50-60%), Carbon Dioxide (30-40%), Hydrogen (5-10%), H_2S and Nitrogen (Traces).**
80. Conversion of Nitrates to atmospheric nitrogen is known as **Denitrification.**
81. Surface to volume ratio in bacteria is **High.**
82. Conversion of atmosphere nitrogen gas into ammonia by micro organisms is due an Enzyme **Nitrogenase.**
83. Example of Sulfur oxidizing bacterium ***Thiobacillus thiooxidans.***
84. Cell wall lacking bacterium example **Mycoplasma.**
85. Bacteria form endorpores when they are in **Stationary** phase of growth curve.
86. Enzymes that are always produced by the cells are **constitutive enzymes.**

87. Enzymes that are always present the cells are **Inducible enzymes.**
88. Pasteurization of milk under low temperature holding method is done at **62.8°C for 30 minutes** followed by rapid cooling.
89. The conversion of organic compounds into inorganic compounds and into their constituent elements known as **Mineralization.**
90. The methanogenic bacteria which utilize **acetate, CO_2 and H_2** to produce methane (Bio gas).
91. Gelatin changes from gel to liquid at a temperature of about **28°C** origin of life from other living organism called as **Biogenesis**.
92. Development of 'Enrichment culture technique' by **Beijerinck and Winogradsky.**
93. Father of chemotherapy **Paul Ehrlich**.
94. **Hellriegel and H. Wilfarth** showed the symbiotic relationship between bacteria and leguminous plants.
95. The term Mycorrhiza discovered by **Frank**.
96. Coined the term 'Rhizosphere' **Hiltner** phyllosphere term coined by **Ruienen**.
97. Example of Extreme thermophiles ***Pyrodictum*.**
98. Example of Microaerophilic bacteria ***Azospirillum*.**
99. Conjugation discovered by **Beedle and Tatum**.
100. Bacteriophage reported by **Twort and Herelle**.
101. Rhizobium can fix **150-200 kg N/ha/yr.**
102. BGA(Blue green algae) can fix **35 Kg/N/ha/**season.
103. Fastest N_2 fixing plant ***Sesbania rostrata*.**
104. Actinomycetes can fix **50-150 kg N/ha/yr.**
105. Microflora associated with plant roots are called **Rhizosphere**.
106. The plating method which is useful in easy isolation of mutants is **Replica plating.**
107. *Spirulina* used as dietary supplement belongs to the microbial group **Cyanobacteria (Blue-Green Algae)**.
108. The chemical nature of Agar-Agar is **Polysaccharide.**
109. Endoflagellum present in **Spirochets.**
110. Transfer of genetic material through bacteriophage in bacteria is **Transduction.**
111. Capsid is made up of subunits known as **Capsomeres.**

112. Biological specificity of disease agent was demonstrated by the scientist **Robert Koch.**
113. Viroids contains only **Nucleic acid.**
114. Minimum water activity required for most food spoilage bacteria **0.91 aw.**
115. Butanediol type of fermentation is carried out by bacterial species **Enterobacter.**
116. Fungal insecticides which controls the potato beetle ***Beauveria bassiana.***
117. Microorganisms which derive their energy from oxidizing inorganic compound are **Chemoautotrophas.**
118. Citric acid is produced by ***Aspergillus niger.***
119. **Submerged culture** means Culturing the organism is grown in a liquid medium with vigours aeration and agitation.
120. Methane percentage in biogas **50–60%.**
121. Medium used in confirmation test of coliforms in water is **Eosine Methylene blue.**
122. VAM fungi will contributes to the uptake of **P** (Phosphorus).
123. The tetrameric protein binds to RNA polymerase and promotes its termination is **Rho factor.**
124. Substitution of purine by pyramidine is called as Transversion.
125. The key enzyme which is responsible in the EDP pathway is **2-keto – 3 deoxy gluconophosphate aldolase (KDPG).**
126. The complete set of gene present in an organism is called **Genome.**
127. The units of DNA which moved from one DNA molecule to another inserting themselves are called **Transposons.**
128. When gram negative bacteria treated with lysozyme they form **Sphaeroplast.**
129. Spirochetes will form a special type of flagella called as **Endoflagella.**
130. Lactic acid fermentation carried out by ***Lactobacillus***
131. Examples of Cyanobacteria ***Nostoc, Anaebenae.***
132. Streptomycin antibiotic produced by ***Streptomyces griseus.***
133. Example of Bacterial insecticides ***Bacillus.***
134. Example of Viral insecticides **Nuclear poly hedrosis virus.**

135. Viruses multiply inside the host cell by **Replication.**
136. ***Bacillus thuringiensis*** is used as Bioinsecticide.
137. An oxygenic photosynthetic bacteria absorb light in the range of **720-1030 nm.**
138. Bacterial chromosome consists of **Double standard DNA.**
139. The formation of ATP by phosphoryl group transfer from a substrate (1, 3, biphosphoglycerate) is called **Substrate level phosphorylation.**
140. The site of attachment of RNA polymerase in "Lac" operon is **Promoter.**
141. Human Insulin is produced from a recombinant strain of ***Escherichia.***
142. Symbiotic nitrogen fixing bacteria is ***Rhizobium.***
143. Butane diol type of fermentation is carried out by ***Enterobacter*** **sp.**
144. Purple sulphur bacteria are example of **Chemo autotrophs.**
145. **Exomicrobiology** study of exploration of life in outer space.
146. **Geo Chemical Microbiology** extracting metals from low-grade ores through microbial activities.
147. ***Thiobacillus thiooxidans & T. ferroxidan*** capable of recovery of copper & uranium from low grade ores of these metals.
148. **Mutation** always change the genotype & such change may or may not be expressed in phenotype.
149. Spontaneous mutation occurs at the rate $\mathbf{10^{-8}}$.
150. **Mutation rate:** Rate of mutation is the probability that a gene/cell mutate at any cell division or the average number of mutation /per cell per generation = 10^{-8} / Cell / Generation.
151. **Mutation frequency**: proportion of mutant or no. of mutant cell contained in the bacterial population at any time.
152. Mutation frequency is higher than mutation rate 10^{-4} to 10^{-8}. The rate of back mutation is 10^{-8} per cell division.
153. Poly auxotroph's can be used as stable markers in the study of **bacterial genetics.**
154. **Nitrites & Nitrosoguanidine:** induce change in DNA through alkalization of the bases of DNA, causes mutation of high frequency.
155. **Bromouracil:** substitute thymine in DNA (point mutation).
156. **Aminopurine:** substitute adenine in DNA (point mutation).

157. **Acridine dyes:** insertion & deletion of 1 or few base pairs in DNA (frame shift mutation).
158. **Nitrous acid (N_2O) & Nitrates (NO_3):** cause deamination of Nitrogen bases.
159. **Ionizing radiations:** Eg: X-rays, Gamma rays.
160. **Non-ionizing radiations**: Eg. UV rays.
161. **UV rays:** formation of pyrimidine dimmers & pyrimidine hydrates.
162. The injury of DNA caused by UV radiation is repaired immediately by **photo irradiation** with visible light of 300-400 nm lambdas. The mechanism is known as **photo reactivation.**
163. **X-ray, Y–rays, and Neutrons:** cause translocation or inverse of bacterial genome, ionize and break molecules in cells forming radicals. (sparsely ionizing).
164. **Plasmid :** is a self-replicating, extra chromosomal hereditary circular DNA found in certain bacteria and fungi.
165. **Episome:** it is a plasmid that can integrate reversibly with the characteristics of its bacteria host & behaves as part of chromosome & is able to multiply independently of chromosome.
166. The first plasmid discovered was **F - plasmid** in *E. coli.*
167. This combining of genes from two different cells c/as **recombination.**
168. **Conjugation: by Beadle & Tatum Lederberg** in ***E. coli.***
169. **Conjugation** :The donor (F^+) cell comes in physical contact with the recipient (F^-) cell via Conjugation Bridge formed by sex pili & transfers its genome to recipient cell.
170. **Transduction: by Zinder & lederberg** in ***Salmonella* sp.**
171. **Transduction:** A portion of DNA from one bacterium (donor) is transferred by a bacteriophage (Vector) to another recipient cell.
172. Lytic phage = virulence = T phage.
173. Lysogenic phase = Temperate = lambda phage.
174. **Transformation: By F. Griffith** in ***Streptococcus pneumonia.***
175. **Transformation:** Process where by cell free or naked DNA containing limited amount of genetic information is transferred from one bacterial cell to another.
176. Bacteria ***Zymomonas mobilis*** employ the ED **Entner- Douderoff** pathway.

177. **Entner- Douderoff pathway** Glucose = 2 Ethanol + 2 CO_2 + 1 ATP.
178. Alcohol productivity of ***zymomonas*** is higher than yeast because of this fermentative pathway.

BIOFERTILIZERS: Bio fertilizers are microbial inoculants or carrier based preparation containing living or latent cells of efficient strains of nitrogen fixing, Phosphate solubilizing & cellulose decomposing micro organisms.

Biofertilizers include

1. Symbiotic N_2 fixers Ex: ***Rhizobium***.
2. Non-symbiotic free living N_2 fixer ex. ***Azotobacter, Azospirillum***.
3. BGA – Inoculant *ex.* ***Azolla – Anabacna***.
4. Phosphate solubilizing Biofertilizers, Ex: ***Bacillus, Aspergillus, Penicillium***.
5. Phosphate absorbing Biofertilizers, Ex: ***VAM fungi, Glomus* sp. & *Aculospora* sp.**
6. **Mycorrhiza:** interaction between roots of higher plant & fungi
7. **Nobbe & Hiltner** produced the 1st **Rhizobium** bio fertilizer under the brand name 'Nitrogen' for 17 different legumes.
8. Permissible contaminants level in Biofertilizer are No contamination at 10^6 dilution (PSB), 10^7 dilunation (***Azotabacter, Azospirillum***), 10^8 dilution (***Rhizobium***).
9. Biofertilizer should contain minimum of **10^8 viable cells** of the carrier material on dry weight basis.
10. Important organsims for **Sugarcane** set inoculation: ***Azotabcter or Azospirillum***.
11. **P** is often limiting factor for growth & N_2 fixation by **BGA**.
12. ***Acetobacter*** is endophytic N_2 fixer in sugar rich crop Ex: **Sugar cane & Sugar beet**.

Food Microbiology

- **Spoilage** means decay or decomposition of an undesirable nature.
- Degradation process responsible for spoilage of food.
- **Putrefaction**: degradation of protein food (Proteolytic) by anaerobic degradation.
- ***Alcaligenes***: Lipolytic degradation.
- ***Clostridium***: Saccharolytic degradation.

- ***Bacillus***: proteolytic degradation
- ***Erwinia***: pectinolytic degradation

(*i*) **Fermentation**: oxidation-reduction process in which an organic compounds i.e., a substrate is partially oxidized or partially reduced.

(*ii*) **Rancidity**: Hydrolytic degradation of fatty acid foods.

- ***Hansenula, Kloeckera and Brettanomyces*** are yeast significant in food microbiology
- Water activity of a food designated by **a_w**

Organisms	**Minimum a_w**
Most spoilage bacteria	0.91
Most spoilage yeasts	0.88
Most spoilage molds	0.80
Halophilic bacteria	0.75
Xerophilic molds	0.65

- Most of the bacteria grow at a pH of **6.8 - 7 (Neutral).**
- Most of the fungi grow at a pH **Acidic Nature (3-4).**
- Yeasts, which can grow at value below **3.5.**
- The oxidation reduction potential of a system is expressed by **Eh.**

Food	**Eh values**
Plant juices	+300 to +400 mv
Solid meats	around –200 mv

- Antimicrobial compound present in milk is **lactenin.**
- Antimicrobial compound present in egg white is **Lysozyme.**
- **Psychrotrophs** belong to the genera *Alcaligenes, Pseudomonas* and *Streptococcus.*
- These **Psychrotrophs** organisms grow well at refrigerated temperatures.
- Most thermophilic bacteria that grow on foods belong to the genera ***Bacillus*** and ***Clostridium.***
- Molds effected nuts are the major sources of **Aflatoxins.**
- Microorganisms causes discolouration of food by producing pigments.

  Red bread → *Serratiama rcescens*

  Grey mold rot → *Biotrytis cinera*

Black mold rot → *Aspergillus niger*

Green rot → *Pseudonmonas fluorescens*

Black rot in egg → *Proteus*

Pink syrup → *Micrococcus eroseus*

- Principles of food preservation:

  Microbiostatic means Inhibition of microbial growth & metabolism.

  Microbicidal means Killing of Microorgansims.

  Pasteurization:

  (*i*) LTHT = 62.8°C (145°F) for 30 min

  (*ii*) HTST (Flash pasterurization) = 71.7°C (161°) for 15 sec

  (*iii*) UHT = 137.8° (280°F) for 1-2 sec.

- **Phosphatase test**: 'Phosphatase' is a thermoliable enzyme present in raw milk.

  This test used to determine the effecting of pasteurization of milk.

- **Canning**: most successful method of food preservation.

  Developed by **N. Appert**, hence it is also called as appertization.

  N. Appert is called as father of canning

- Temp. for canning 100°C for high acid food.
- Temp. for canning 121°C for low acid food.

  Canned Food contain *Clostridium botulinum* causing food poisoning botulism.

  Immersing leafy vegetables in hot water before refrigeration is called **Blanching**.

(*iii*) Use of low temp:

(*a*) **Chilling**: just above freezing point, ex refrigerator

Retards but doesn't prevent multiplication of all **O**[s]

(*b*) **Freezing:**

1. Quick freezing: 30 min 18°C to 34°C Produces small i.e. crystal & cause less damage to food.
2. Slow Freezing: 3 to 72 hr- 15°C Produce large ice crystal & cause more damage to food.

Dehydration: The **removal of water** by drying in the **sun** and **air** or with applied **heat** causes dehydration.

- Product containing moisture below 10% can be stored for longer period under dry condition.
- The preservative effect of dehydration is due to 'Microbiostasis' (not necessary killed).

  Osmotic pressure: Microorganisms are plasmolysed

  (*a*) **Brine solution:** salt concentration of 20-30%.

  (*b*) Concentrated sugar: use to preserve Jam, Jelly and Honey.

  Osmophilic' yeast can grow in honey & produce sufficient CO_2 to burst the jar.

(*iv*) **Chemicals:**Cause Microbiostatic & Microbiocidal Action

- **Benzoic acid & Benzoate** used to preserve high acid food.
- **Salicylic acid & salicylate** used to preserve Fruits & Vegetables.
- **Smoking** used to preserve Fish & Meat.
- **Boric acid** used to preserve Butter (antibacterial).
- **Sulphorous acid & sulphites** used to preserve Wines.
- **Sodium diacelate, AA, LA propionic acid** used to preserve Breads.
- **Sodium nitrate & nitrate** used to preserve Curing of meat & fish.
- **Ethylene oxide** used to preserve spices.

(*v*) Radiation:

(*a*) **NonIonizing radiation**: limited penetration, control Microorganisms on surface only Eg: UV–radiation.

(*b*) **Ionizing radiation**: kill deep seated microbes Ex: X-rays, Gamma rays.

- **Food Poisoning:** Classification of food borne diseases are suggested by **F.L. Bryan.**
- Toxicity introduced into food by microorganisms an their toxins

  (*i*) Food intoxication → Caused due to consumption of food contaminated toxic product of Microorganism.

  (*ii*) Food infection → ingestion of contaminated food.

(*i*) **Food intoxication:**

- Caused due to consumption of food contaminated C toxic product of MO[s].

  Ex. Botulism

  Botulism caused by *Clostriduim botulinum* (G+ve spore forming anaerobe).

- *Clostriduim botulinum*'**Exotoxins**' which is a **protein** causes **Botulism**.

(*b*) Common food poisoning ***Staphylococus aureus***.

- *S. aureus* produces an exotoxin as **'Enterotoxin'**

(*ii*) **Food infection:**

- Illness caused by entrance of the bacterium into body through ingestion of contaminated food & the toxin of host body to their presence or their metabolites.

(*a*) **Salmonellosis:** food poising by secreting **Endotoxin**.

Mycotoxins: poisonous compound produced by **molds** in staple food.

Ex: **Afllatoxin** produced by **A. flavus** causes Turkey – X disease.

Amanitin produced by ***Amanita musearea*** (Mushroom).

Ergotin, Ergometrin produced by ***Claviceps purpurea*** (Ergot of rye).

- G +ve organism produces **Exotoxin**.
- G –ve organism produces **Endotoxin**.
- Presence of **20 ppm of aflatoxin** in food is max. permissible level.
- **Poliomyelitis (PMV), jaundice (hepatitis v)** is food borne viral infection.
- Storm fermentation : a rapid fermentation of lactose in milk due to excess gas production by ***Clostridium perfringea***.
- Thermoduric bacteria **Ex.** ***Microbacteria, Micrococcus, B. subtilis, S. thermophilus***.
- Mastitis caused by ***S. aureus*** is transmitted by human to cow.
- Sourcing of milk is brought by lactic acid bacteria like ***S. lactis, S. cremoris & Lactobacilli***.
- Propionibacterium is exclusively used in the preparation of **Swiss cheese**.
- Due to surface growth of pigmented bacteria or mold

Blackish, Brown, Blue → ***Pseudomonas* sp.**

Blue colour → ***S. lactis***

Yellow colour → ***Flavobacterium***

Red colour → ***Serratia marcescens, Brevibacter***

Pink colour → ***Micrococcus, Torula (Yeast)***

Flavour production:

Malty flavour → ***S. lactis var. maltigens***.

Fishy flavour → ***P. icthyosmia***.

Sp. of ***Lactobacillus & Streptococcus*** are commonly used as starter culture.

Yogurt (Dahi) → ***S. lactis, L. bulgaricu, S. thermophilles***.

- **Roquefort** ripened by ***Penicilluim roqueforti*** is semi soft type of cheese.
- **Swisscheese** is **hard** type of cheese.
- **Cheese** with characteristic eyes **Swiss cheese**.
- *Mucor* is involved in the making of Tofu.
- Bitter taste in milk due to **Proteolytic bacteria**.
- Rancidity is related to **Lipid spoilage**.
- **Sauerkraut** : fermented cabbage in brine.
- Steps in Cheese manufacture: curdling–draining—salting—ripening.
- **Vinegar** exhibits dual type of Fermentation.
- Pure cucumber pickles also called as **Dill pickles**.
- **Tempeh:** solid food prepared with Soyabean processed with sps of ***Rhizopus***.
- **Sufu:** Chinese cheese prepared by soyabeans with ***Actinomuco relegans & mucor***.
- **Miso** :prepared by soyabeans and rice ***Aspergillus oryzae***.
- Soft rot of vegetables causative organism ***Erwinia caratovera***.
- Gray mold rot caused by ***Botrytis cinerea***.
- Blue mold rot caused by ***Pencillium digitatum***.
- Black mold rot caused by ***Aspergillus niger***.
- Pink mold rot caused by ***Tricothecium rosem***.
- Watery soft rot caused by ***Sclerotinias clerotium***.
- Ropymilk caused by ***Alkaligenes viscolatics, Micrococcusredden*** rechii, *Klebsiellaoxtoka*
- Alkaline milk caused by ***Alkaligenes viscolatics, Pseudomonas fluorescens***.
- Sour milk caused by ***Streptococslactis, Lueconostoc, Clostridium***.

- Bitter milk caused by proteolyticbacteria, ***some Coliforms, Asporogenous yeasts...***
- Blue milk caused by ***Pseudomonas syncyanea***.
- Yellow milk caused by ***Pseudomonas synxantha, Flavobactrium***.
- Red milk caused by ***Serratia, Brevibacterium erythrogenes***.
- Brown milk caused by ***Pseudomanas putrefaciens, Ps. fluorescens***.
- Benzoic acid maximum permissible concentration **0.1%**.
- Methyl paraben maximum permissible concentration **0.1%**.
- Sodium nitrate maximum permissible concentration **500 ppm**.
- Sodium nitrite maximum permissible concentration **200 ppm**.
- Ethylene oxide maximum permissible concentration **50 ppm**.
- Example of Photoautotrophic algae ***Cynobacteria*** (micro algae).
- Example of Photoautotrophic Green sulfur bacteria ***Chlorobium, Chloropseudomanas***.
- Example of Photoautotrophic Purple sulfur bacteria ***Chromatium, Thispirillum***.
- Purple sulfur bacteria and Green sulfur bacteriaposses a special photosynthetic pigment called as **Bacteriochorophyll**.
- Concept of Chemoautotrophy was formulated by **Winogradsky** studying in the organism **Beggiatoa**.
- Example of filamentous gliding mixotrophy sulfur oxidizing bacteria **Beggiatoa**.
- Example of Chemoautotrophic sulfur oxidizing bacteria ***Thiobacillus thioparus, Thiobacillus intermedius***.
- Non sulfur purple bacteria ***Rhodomicrobium, Rhodospirillum***.
- Example of Chemoheterotrophic Nitrifying bacteria is ***Nitrobacter***.
- Example of photo heterotrophic purple non sulfur bacteria ***Rhodopseudomonas***.

Rhodospirillum

- Example of Hydrogen oxidizing bacteria ***Alkanigenes, Aquaspiriumm***.
- Reduction of oxygen with hydrogen this reaction is called as ***Knallgas reaction***.
- Example of chemoorganoheterotrophs **Protozoa, fungi (slime moulds), viruses, diatoms**.

Industrial Microbiology

- Fermentation concept developed by **Louis Pastuer.**
- Cell free Fermentation developed by **Buchner.**
- **Solid state fermentation** means growth of culture without presence of added free water.
- **Molasses** serve as a source of **carbon and energy.**
- **Corn steep liquor** source of **Nitrogen.**
- **Siliconesand vegetable oils** serve as **antifoam agents.**
- **Whey** is a fermentation medium that comes from **diary industry**.
- ***Pencillium notatum*** culture was significantly improved for better yields in the early days of fermentation technology.
- **Impeller** used for agitation in a **Biareactor.**
- **Turblence** is due to presence of **Baffels.**
- **Sparger** is aerating device in bioreactor.
- Quick and widely used method to determine cell content in a bioreactor **Packed Cell Volume (PCV).**
- **Fluidized bed reactor:** microbes are cultured as biofilms on surface of suspended solid particles.
- **Bubble cap fermentor**: cultivation of microbial cells using gaseous substrates.
- Growth along a central streak to test inhibition of test organisms is called **Giant colony technique**.
- Photosynthetic organisms mostly used in the **biofuel** production.
- **Screening** is the process of selecting industrial cultures.
- Identification & isolation of potential culturesfromnatral samples such as air, water and soil called as **Primary Screening.**
- **Downstream processing** is harvest and recovery of industry prodct.
- Toxic effect of Heavy metals on microbial cells referredas **Oligodynamic action**.
- **Bioreactor** is the vessel in which microbes and their metabolites are used to produce viable products.
- Undistilled beverages **Ex : beer, wine also called as soft liquor**.
- **Soft liquor** contain lower percentage of alcohol up to **20%**.
- Distilled beverages **Ex: whisky, rum, gin, brandy, vodka, fenny** also called as **hard-liquor**.

- **Distilled beverages** contain higher percentage of alcohol more than **20%**.
- Undistilled product of grain mash fermentation brought by yeast known as **Beer**.
- In beer making **mashing** causes hydrolysis of malt starch.
- Hops in beer contribute to **antiseptic property, flavor and colour**.
- *Sacharomyces cervisiae* (brewer's yeast) & Bottom yeast used in **Beer** production.
- Brewer's yeast also called as ***Sacharomyces cervisiae***.
- Top fermenter (yeast) organism in wine production ***Sacharomycescervisiae***.
- Bottom fermenterorganism ***Sacharomyces cervisiae*** **and** ***Clostridium*** **sp.**
- **Pitching** means addition of inoculum to fermentation medium.
- **Wine** is the undistilled product of fruit juice fermentation by yeast.
- ***Sacharomyces cervisiaevar, ellipsoidens*** mostly used in wine fermrntation.
- Crushed fruits with juice are called as **must**.
- Must treated with $\mathbf{SO_2}$ to prevent microbial spoilage.
- **Still wines** do not contain CO_2 Ex: sherries, vermouth.
- **Sake** is Japanese alcoholic beverage prepared by fermenting rice.
- **Kuass:** Russian product prepared by barly mart, rye malt.
- **Pulque:** produced by using Agave juice.
- **Kuass, chicha & jaette are** undistilled alcoholoic beverages.
- **Vodka** produced in Russia by yeasty fermentation of potatoes.
- **Rum:** distilled fermented juice of sugarcane.
- Rum contain **40% of Alcohol**.
- **Gin:** fermented mash or raw grain.
- **Brandy:** distilled product of fermented grape juice.
- Brandy contain **40-50% of Alcohol**.
- **Fenny:** Fermented juice of cashew apple.
- Fenny also called as **Cashew apple liquor** in Goa.
- Citric acid production first developed by **Wehmer**.
- Citric acid producing organism ***Aspergillus*** sp. (black and white mold).

- Acetic acid producing organism ***Acetobacter*** sp.
- **Molasses** used as media in the production of **Alochol.**
- **Wheyused** as media in the production of **lactic acid.**
- **Soymeal** used as media in the production of **streptomycin.**
- **Corn steep liquor** used as media in the production of **pencillium.**
- ***Torulautilis*** type yeast used in the animal feed.
- Production of alkaloids – ***Claviceps perpurea*.**
- ***Xanthomonas*** organism used in production of polysaccharides as food additives.
- Example of **Fumaric acid** producing organisms *Rhizopusnigricans, Mucor, Penicillium.*
- Amylase, Pectinase, Protease → ***Aspergillus* sp. *Bacillus subtilis*.**
- Cellulase produced from ***Trichoderma viride*.**
- ***Curvlaria lunata*** used in the production of steroid prednisolone.
- ***Rhizopus nigricans* & *Arthobacter*** used in Steriods biotransformations.
- **Bacitracin** produced from ***Bacillus licheniformis*.**
- Streptomyces Venezuela is used in the production of antibiotic **Chloramphenicol**.
- ***Pseudomonas*** sp. employed in environmental management.
- **Invertase** used in producing artificial honey.
- Amino acid valine is produced from ***Brevibacterium***.
- Antifngal antibiotic produced Griseofulvin is produced by ***Pencillium.*** sp.
- Majority of harmones (Insulin) are produced industrially from → ***E.coli*.**
- Bakery product → ***Candida milleloti*** (Bakers yeast).
- Single Cell Protein (SCP) → Algal: ***Spirulina, Scenedesmus, Chlorella*.**
- Bacterial: ***Cellulomonas, Pseudomonas*.**
- Fungi: ***Aspergillus oryzae, Candida, Fusarium moniliformae*** and ***saccharomyces cerevisae*.**
- Flemming → **fungi produces antibiotics.**

OOO

CHAPTER 4

Soil Science and Agricultural Chemistry

1. Science that deals with soil genesis and classification is called as **Soil Pedology.**
2. Study of the soil from the stand point of higher plants called as **Edaphology.**
3. Inherent capacity of soil to supply the plant nutrients in adequate amount without any toxicity to the crop plants **Soil fertility.**
4. The word soil is derived from latin word **Solum.**
5. Dominant element in the earth crust **Oxygen.**
6. Second most dominant element earth crust **Silica.**
7. The most abundant metallic element **Aluminum (Al).**
8. Mineral soils hardly contains organic matter **1-2%.**
9. Density of the earth **5.5 g/cm^3.**
10. Percentage of CO_2 in the atmosphere **0.03% (300 ppm).**
11. Study of rocks is known **Petrology.**
12. Study of description of rocks are called **Petrography.**
13. Study and origin of rocks are called **Petrogenesis.**
14. The cooling down of molten magma which rocks are formed **Igneous rocks.**
15. The predominant rock in the composition of the upper 5 km of earth crust rocks are known as **sedimentary rocks.**
16. Rocks are formed from igneous rocks by the influence of high pressure and temperature called **Metamorphic rocks.**
17. The silica content in acid igneous rocks is **65%.**

18. The silica content in Basic igneous rocks is **50%**.
19. Granite, Basalt are which type of rocks **Ingenues rocks.**
20. Example for acid igneous rock **Granite.**
21. Example for Basic igneous rock **Basalt.**
22. Igneous rock are **Granite, Basalt and pumice.**
23. Sedimentary rocks are **Limestone, Dolomite and Sandstone.**
24. Metamorphic rocks are **Marble, genesis, schist, slate and graphite.**
25. Stratification is most common feature in **sedimentary rocks.**
26. Mechanically formed rocks are **Sand stone shale.**
27. Organically/ Biochemically formed sedimentary rocks are **Lignite and Anthracite.**
28. Inorganically formed rocks by the process of evaporation **Gypsum.**
29. Sedimentary rocks are formed by the process of precipitation & flocculation **Limestone and dolomite.**
30. Sandstone on metamorphosis is converted to **Quartzite.**
31. Lime stone on metamorphosis is converted to **Marble.**
32. Shale stone on metamorphosis is converted to **Slate.**
33. Coal stone on metamorphosis is converted to **Graphite.**
34. Volcanic rock that is in vesicular structure **Pumice.**
35. An example of floating rock **Pumice.**
36. Weathering is a **geological process & destructive** process.
37. Sandstone are eventually converted to quartzite due to **Hydro metamorphism (water).**
38. Limestone are eventually converted to quartzite due to marble due to **Thermo metamorphism (Temperature).**
39. Granite are eventually converted to Genesis due to **Dynamo metamorphism.**
40. Genesis are eventually converted to schist due to **Dynamo metamorphism.**
41. The primary mineral which form the chief constitutents are called as **Essential mineral.**
42. Primary mineral occur in small quantities are called **Accessory mineral.**
43. First phase of soil formation is **Weathering.**
44. Soil formation from parent material done by **Pedogenic process.**

45. Predominant primary mineral present in igneous rocks is **Feldspar**.
46. The nutrient element present in orthoclase and microcline is **K**.
47. An example of sodium feldspar **Albite**.
48. An example of potassium feldspar **Anorthite**.
49. The color of biotite is **Black**.
50. The color of muscovite **White**.
51. Mineral which one is most resistant to weathering **Quartz**.
52. Mineral which one is least resistant weathering **Olivine**.
53. Mineral which one is more weather able capacity **Olivine**.
54. Light minerals are the ones which have specific gravity **<2.85g/cm^3**.
55. The property of hardness of mineral measured by **Mohr's scale**.
56. Chief constituents of sandy fraction **Quartz**.
57. Most dominant mineral in the earth crust **Feldspar (48%)**.
58. B containing mineral is **Tourmaline**.
59. Major source of Mg **Dolomite**.
60. Major source of P **Apatite**.
61. Major source of Mo **Olivine**.
62. Major source of K **Orthoclase**.
63. Specific gravity of heavy minerals **>2.85 g/cm^3**.
64. Specific gravity of low minerals **<2.85 g/cm^3**.
65. Physical weathering of rocks is process of **Disintegration**.
66. Chemical weathering of rocks is process of **Decomposition**.
67. Sand stone contains 78.64% **SiO_2**.
68. Lime stone contains 42.74% **CaO**.
69. The chemical weathering process is dominant in ill drained condition **Reduction**.
70. The rock that on weathering forms block soil **Basalt**.
71. Peeling away of surface mass of rock due to variation in temperature is **Exfoliation**.
72. The weathering process involved in the formation of limonite from hematite **Hydration**.
73. The important chemical weathering process is **Hydrolysis**.
74. The weathering process involved in the formation of kaolinite from biotitic **Hydrolysis**.

75. The end product of weathering **Parent material.**
76. Soil formation factors initially developed by **Dokuchaev.**
77. Jenny equation of soil formation **S = f (U, B, R, P, T).**
78. Active factors of soil formation **Climate, Biosphere.**
79. Passive factors of soil formation **Parent material relief, time.**
80. Active and passive soil formation given by **Joffe.**
81. Strictly speaking there is no biological weathering **Jofee.**
82. **Jenny** defines parent material is initial stage of soil system.
83. The soil forming factor that has no relevance after soil reaches its maturity **Time.**
84. The parent material that is transported by water **Alluvium.**
85. The parent material that is transported by wind **Aeolian.**
86. The parent material that is transported by gravity **Colluvium.**
87. Silt material transported by wind called **Loess.**
88. Sand material transported by wind called **Aeolian.**
89. The parent material deposits by ice **Moraine.**
90. High topographic positions soil formed **Red soil.**
91. Low topographic positions soil formed **Black soil.**
92. Fundamental soil farming process **Humification, Illuviation, Elluviation.**
93. Most important factor for soil formation **Climate.**
94. Five weathering stages given by **Mohr vanbaren.**
95. The process of removal of material from upper horizons of profile to lower horizon called **Elluviation /mitigation.**
96. The process of deposition of soil materials in the lower layers is termed as **Illuviation.**
97. Removal of silica and accumulation of sesquioxides called as **Laterization.**
98. Removal sesquioxides and accumulation of silica called as **Podzolization.**
99. The process of darkening of surface soil **Melanization.**
100. The process of soil formation under poor drainage conditions called **Gleization.**
101. Soil formed by the process of gleization called as **Hydromophic soils.**

102. The process of mixing of the soils **Pedoturbation.**
103. Mixing of materials in solum by the churning process caused by swell shrink **Agrillic pedoturbation.**
104. The vertical section of soil through all its horizon and it extends up to its parent material **Soil profile.**
105. Smallest volume that can be recognized as a soil **Pedon.**
106. Soil solum **A + E + B.**
107. Regolith **A + B + C.**
108. Father of soil science, gave the factors of soil formation **Dokuchaiev.**
109. The diagnostic surface horizon called as **Epipedon.**
110. The diagnostic sub surface horizon called as **Endopedon.**
111. An epipedon formed due to long continued system of farming **Anthropic epipedon.**
112. The epipedon with base saturation is 50% belongs to **Mollic epipedon.**
113. The epipedon base saturation less than is 50% belongs to **Umbric epipedon.**
114. Argillic subsurface diagnostic horizon is diagnostic feature of **Red soils.**
115. Honey comb structure is a diagnostic feature of **Lateritic soil.**
116. Black soil exhibiting deep cracks when dry belong to the soil order **Vertisols.**
117. Best Agricultural soils in the world are **Mollisols.**
118. The most important soils for agriculture **Alluvial soils.**
119. The highest category in soil taxonomy **Soil order.**
120. The lowest category in soil taxonomy **Soil series.**
121. Soil order that is dominantly organic in nature **Histosol.**
122. The soil under heath or pine vegetations developed into **Spodosols.**
123. Soil under grassland vegetation develop into **Mollisols.**
124. Genetic system of soil classification given by **Dokuchaier.**
125. Morphogenetic system of soil classification given by **Marbut.**
126. Diagnostic horizons are used for the classification of soils **Great soil group.**
127. The order that have not been observed in India **Gelisols.**
128. The soils do not exhibit any elevation and illuviation is **Black soils.**
129. The largest are in India occupies by **Alluvial soils.**

130. The characteristics feature of soil crusting observed in **Red soil.**
131. The characteristics feature of deep cracks observed in **Black soil.**
132. The major constraint of alluvial soils is **Stratification.**
133. Desert soils are comes under **Aridisols.**
134. Saline soils are also called **White alkali.**
135. Saline soils are reclaimed by **Leaching.**
136. Alkali soils are reclaimed by **Gypsum applying.**
137. Acid soils are reclaimed by **Liming.**
138. Red soil dominating state **Tamil Nadu.**
139. Black soil dominating state **Maharashtra.**
140. Which clay mineral is rich in red soil **Kaolinite.**
141. The most fertile soils are **Alluvial soils.**
142. Which is one example of 1:1 type clay mineral **Kaolinite.**
143. Which is one example of 2:1 type clay mineral **Montmorillinite.**
144. Which is one example of 2:1:1 type clay mineral **Chlorite.**
145. Amorphus clay mineral found in andosols **Allophane.**
146. Major forest area is surveyed by **Reconnaissance survey.**
147. Cadastral maps of detailed soil survey is **1:5000 to 1:8000.**
148. The optimimum physical condition of the soil for crop growth is called **Soil tilth.**
149. Surface accumulation of organic matter called **Duff/Mor.**
150. Highly weathered soils with an oxic horizon **Oxisols.**
151. Soils with swell shrink type clays having 30% clay **Vertisols.**
152. Recent soil with no diagnostic horizon **Entisol.**
153. The process by which the dispressed clay particles are migrated from upper to lower horizon **Agrilliation.**
154. Removal of clay particularly the fine clay in suspension from the upper part of soil profile **Lessivage.**
155. The relative proportion of soil particles called **Soil texture.**
156. The most permanent physical property in soils **Texture.**
157. The USDA system diameter of silt particles **0.2 to 0.002 mm.**
158. The USDA system diameter of clay soil particles **0.002 mm.**
159. Most commonly used dispersing agent used in mechanical analysit of soil **Sodium hexa meta phosphate.**
160. The value of particle density is in stocks law **$2.65 g/cm^3$.**
161. Soil texture can be done by **International pipette method.**

162. Which one is called as Flesh of soil **Clay.**
163. The weight of a hectre of surface soil **2.24×10^6 kg/ha.**
164. Heavy soils are difficult to bring to proper tilth because of their **Plasticity.**
165. The attractive force between soil solid particle **Cohesion.**
166. The force with which water is held to the soil particles is called **Adhesion.**
167. The arrangement of primary particles and their aggregates called **Soil structure.**
168. Soil structure described into **Grade class type.**
169. When soil structure is suitable for gets are called **Crumb.**
170. When the soils particles more in the horizontal axis than vertical structure is **Platy structure.**
171. Vertical axis is more than horizontal one **Prism structure.**
172. Naturally formed soil aggregates are called **Ped.**
173. Broken peds are known as **Fragment.**
174. The first step in the formation of stable aggregates **Flocculation.**
175. Which element is essential for flocculation **Calcium.**
176. Which element is essential for defloculation **Na.**
177. Synthetic chemicals added to the soil to improve the soil structure called **Soil conditioners.**
178. Theory of bonding of clay clay particles was given by **E. J . Russel.**
179. Anion which is responsible for soil aggregation **Phosphate.**
180. Column or structure is mostly found in **Salt affected soils.**
181. The physical property will not change on compaction is called **Particle density.**
182. Particle density of mineral soil is **2.65 gm/cm^3.**
183. Particle density also called as **True density.**
184. Bulk density of fine textured soils **1.1 to 1.3 g cm^3.**
185. Bulk density of coarse textured soils **1.4 – 1.8 g/cm^3.**
186. Bulk density of organic peat soils **0.5 g/cm^3.**
187. The munsel colour notation of the purity of colour **Chroma indicated.**
188. The munsel colour notation of the dominant spectral colour is indicated by **Hue.**

189. The Muncell color chart the brightness of color is denoted by the term **Value.**
190. Structure is denoted in sequence of **Grade class-type.**
191. Porosity percentage pore space is given by **(1-Pb/Ps) X100).**
192. Soil colloid is made up inorganic colloid Clay and organic colloid **Humus.**
193. Water held between 1/3rd – atmosphere called **Gravitational water.**
194. Water held between 15 – atmosphere called **Wilting point.**
195. Water held between 1/3rd to 15 atmosphere called **Available water.**
196. The process of entry of water into soil called **Infiltration.**
197. The downward movement of water is called **Percolation.**
198. In which soil water infiltration is more rapid **Sandy soils.**
199. Removal of excess water from soils known **Drainage.**
200. Saturated conductivity of water is highest in **Sandy soil.**
201. The rate of saturated flow water in soils in the order **Sandy > loam > clay**
202. The rate of unsaturated flow water in soils in the order **clay > loam > Sandy.**
203. The concept of pH given by **Sorenson.**
204. The concept of pF given by **Schofield.**
205. Water retained between field capacity and permanent wilting point **Available water.**
206. The inner film of water around soil particle is held by **Adhesion.**
207. In a saturated soil, the pressure potential is **Positive.**
208. The amount of energy required to evaporate one gram of water **580 cal/gm.**
209. What is the valve of pF scale at fields capacity **2.53.**
210. What is the valve of pF scale at permanent wilting point **4.18.**
211. In the field matric potential of soil is measure by **Tensiometer.**
212. Tensiometer can be used to measure soil water tension up to **0.85 bars.**
213. Soil water potential in laboratory can be measured by **Porus cup apparatuse & pressure plate apparatus.**
214. In neturon probe the slow neutron are observed by **BF_3 Gas.**
215. The attraction of water from the atmosphere by the soil colloids is called **Adsorption.**

216. Thermal conductivity is maximum ion **Sandy soils.**
217. The specific heat of water is **1 cal/g.**
218. Thermal conductivity of soil in different order **Sandy > Loam > clay.**
219. In soils heat is mainly transferred by **Conduction.**
220. Which law is deals with heat conduction in soils **Fourier's law.**
221. Movement of gas is due to concentration gradient is called **Diffusion.**
222. For normal crop growth, aeration porosity of soil must be not less than **20%.**
223. Oxygen diffusion rate value optimum of the crops **$20\text{-}40 \times 10^{-8}$ g/cm^2/ min.**
224. Fundamentals of soil physics written by **Hillel. D.**
225. The Negative logarithm of H+ ions concentrations called as **Soil pH.**
226. CEC of montomorilinite **80-150 meq/100 g soil.**
227. Cation exchange capacity of humus **200 meq/100 g soil.**
228. Fertile soils are saturated with **Ca^{2+} and Mg^{2+} ions.**
229. C:N ratio of the organic matter **10:1.**
230. C:N ratio of the Indian soil **12:1.**
231. The smell of soils after fresh shower is due to ***Actinomycetes*.**
232. Symbiotic N- fixing bacteria is ***Rhizobium*.**
233. Associate symbiosis carries by ***Azosiprillum*.**
234. Nitrifying bacteria as are ***Nitrasomonas and Nitrobactor*.**
235. What is the percentage of CO_2 in biogas **30-40%.**
236. Which biofertilizer is suitable for cereals **Azotobacter/ Azosprillum.**
237. The major biofertilizer is suitable for sugarcane crop **Acetobater.**
238. Both organism are mutually benifited this association called as **Symbiosis.**
239. The ammonia is converted as nitrate **Nitrasomonas.**
240. The Nitrate is converted as nitrite **Nitrobacter.**
241. What is the pH of acid sulphate soils **<3.5.**
242. Saline soils are also called as **Solanchalka.**
243. Central soil salinity research institute located **Karnal.**
244. Example for high salt tolerant crop **Date palm.**
245. Alkali soils are called as **Sodic / solonetz.**

246. The amendment suitable for calcareous sodic soil is **Pyrites.**
247. After reclamation of sodic soils the major nutrient which becomes deficient is **Phosphorus.**
248. Buffering capacity is high for **Humus & clay.**
249. Redox potential of oxidized zone of submered soils is **>+400mv.**
250. Submerged of soil decreases the availability of **S, Cu, Zn.**
251. Crops highly tolerant to chlorides **Apple & apricot.**
252. RSC of irrigation water (MeL^{-1}) **(CO_3^{2-} + HCO^{3-}) – (Ca^{2+} + Mg).**
253. For a good water sample the SAR value **<10.**
254. For a good water sample the EC value **<2 dS/m.**
255. For a good water sample the RSC value **<2.5 meq/L.**
256. The rapid tissue test for determination of K uses **Sodium cobalt.**
257. Quick test for P is **Brays modified test.**
258. Tissue test for NO_3 – N is **Diphenyl ammine test.**
259. The test crop used for Jenny pot culture is **Lettuce.**
260. Microbiological test foe Cu & Mg **Mulders technique.**
261. The formula for radio isotope technique of A value **A = B(1-Y)/Y**
262. The critical value of percent nitrogen in rice crop is **1.4.**
263. Scattering of detached soil particles by the impact of rain drops called **Splash erosion.**
264. Removal by rain of very thin layer of soil from the entire surface called **Sheet erosion.**
265. Universal soil loss equation A = RKLSCP K denotes **Soil erodability.**
266. Dominant type of wind erosion in soil **Saltation.**
267. Eutrophication of lakes caused mainly by **P.**
268. The metal is responsible for itai – itai diseases in human beings **Cd.**
269. The C;N ratio is <20:1 leads to **Mineralization.**
270. The C;N ratio is <30:1 leads to **Immobilization.**
271. The C:N ratio of paddy straw **100:1.**
272. Maximum losses of nutrients occur in which method **Heap method.**
273. Best method of Forage of FYM **Cover pit method.**
274. Materials yielding foul smell of organic compounds are called **Mercaptans.**

275. The fertilizer is added to prevent nitrogen losses during the decomposition of organic matter **SSP.**
276. The composting method which is aerobic and anaerobic fermentation process **Bangalore.**
277. Largest mechanical compost plant in world **Israel.**
278. Night soil digesters developed from grist of bones called **Deamination.**
279. Process of removal of proteins from grist of bones called **Deamination.**
280. Oil cake contain high amount of N **Safflower cake.**
281. The product of sea bird excrement **Guano.**
282. Suitable green manure crop recommended for alkaline / sodic soils **Sesbania aculeate(Dhaincha).**
283. An example of 1:1 type expanding clay mineral **Halloysite.**
284. The C- axis basal spacing of montmorillinite **14 A°.**
285. In montmorillinite negative charges arise mostly by **Isomorphs substitution.**
286. The H bonding is present in the clay mineral **Kaolinite.**
287. Amorphus non crystalline alumnio silicate clay mineral **Allophane.**
288. The range of CEC of montmorillinite **80-120 Col (p+)/ kg.**
289. Potassium fixation is more in **Vermiculite.**
290. The CEC order of clay minerals **vermiculite > Illite > Kaolinite.**
291. The CEC was first reported by **Thomas way.**
292. What is the CEC /AEC ration of kaolinite **0.5.**
293. What is the CEC /AEC ration of montmorillinite **6.7.**
294. The constituent of humus soluble in alkali insoluble in acid **Humic acid.**
295. The constituent of humus both alkali and acid soluble in **Fulvic acid.**
296. The CEC of humus **250-300 meq/100.**
297. The reduction of nitrate into nitrogen called **Denitrification.**
298. The micro organisms which produces antibiotics **Actinomycetes.**
299. The constituent of organic matter resistant to degradation is **Lignin.**
300. In podzolization the dominant microorganism **Fungi.**
301. Conversion of organic form of an element into inorganic form called **Mineralization.**

302. The micro organism involved in oxidation of ammonia to nitrate is **Nitrasomonas.**
303. The bacteria which is fixes atmospheric N **Rhizobium.**
304. Denitrifying facultative anaerobic bacteria **Pseudomonas, Bacillus.**
305. Symbiotic N- fixing actinomycetes **Frankia.**
306. Which enzyme is involved in biological nitrogen fixation **Nitrogenase.**
307. The catalyst used in manufacture of ammonia **Activated iron oxide.**
308. Oldest nitrogen fertilizer manufactured in India **$(NH_4)_2$ SO_4.**
309. By product of coal distillation **Ammonium sulphate.**
310. Which one is neutral fertilizer **CAN.**
311. The first unstable compound produced by the hydrolysis of urea **Ammonium carbonate.**
312. The most suitable N fertilizer for foliar spray is **Urea.**
313. Which is called as nitro lime (or) nitro chalk? **CAN.**
314. The fertilizer induces poor physical condition of soil is **$NaNO_3$.**
315. For water logged rice cultivation the fertilizer that may be avoided is **$NaNO_3$.**
316. An example of soil acting nitrification inhibitor **N-serve / AM.**
317. Most popular phosphorus fertilizer in India _ SSP is **Apatite.**
318. Single super phosphate **Rock phosphate +H_2SO_4.**
319. Triple super phosphate **Rock phosphate +H_3PO_4.**
320. Nitrophosphate formula **Rock phosphate +HNO_3 + NH_3.**
321. Basic slag enriches with phosphors is marketed as **Pelophos.**
322. Muriate of potash is not recommended for **Tobacco crop.**
323. Desliming is important process in manufacturing of **MOP.**
324. Phosphorus fertilizer use efficiency is **15-20%.**
325. Nitrogen fertilizer use efficiency is **25-40%.**
326. Copper sulphate is also known as **Blue vitriol.**
327. Phosphours solubilizing fungi **Aspergillus.**
328. For manufacturing biofertilizers the carrier material used is **Lignite.**
329. The fertilizer guide was written by **H.L.S. Tandon.**
330. Soil consistency and international classification given by **Atterberg.**
331. Law of diffusion given by **Ficks.**
332. Father of Indian soil science **J.W. Leather.**

333. Father of field plot technique **Boussinguault.**
334. The term colloid given by **Thamoas Thomas.**
335. Law of minimum given by **Liebig.**
336. Essentiality of Mo given by **Arnon & Stout.**
337. Essentiality of B given by **Warrington.**
338. A value technique given by **Fried & Dean.**
339. Nutrient mobility concept given by **Bray.**
340. Urea manufacturing process given by **B. Wholer.**
341. Lime requirement is given by **Shoemaker.**
342. Gypsum requirement is given by **Schoonover.**
343. ISSS is located at **Bhopal.**
344. NBSS & LUP located at **Nagpur.**
345. Alkaline permanganate method is used for estimation of **Available N.**
346. Estimation available P in sodic soil the process is **Olsen method (0.5 M Na HCo_3).**
347. In acid soils estimation of available P method is **Bray (0.03 N NH_4 F+ 0.025 N HCl).**
348. The conversation factor to convert percentage P to P_2O_5 **2.29.**
349. Estimation of carbonates the indicator is **Phenolphthalein.**
350. Estimation of Bicarbonates the indicator is **Methyl orange.**
351. Estimation of Ca the indicator is **Muroxide.**
352. Estimation of Mg the indicator is **EBT.**
353. Total digestion for N estimation which prevents loss of $NO_3 - N$ **Salicylic acid.**
354. Indicator used in estimation of organic carbon **Diphinyl amine indicator.**
355. Extract used for determination of K is **Neutral normal ammonium acetate.**
356. The inside of pH meter filled with **0.1 N HCl.**
357. Due to deficiency of Mo causes **Whiptail in cauliflower.**
358. Due to deficiency of Mn causes **Marshy spot of pea.**
359. Phala blight of sugarcane is deficiency of **Mn.**
360. Blossom end rot of tomato is deficiency of **Ca.**
361. Top sickness of tobacco is deficiency of **B.**
362. White bud of maize due to deficiency of **Zn.**

363. Bitter pit of apple due to deficiency of **Ca.**
364. Which elements are called ballast elements **Al & Si.**
365. Which soil order occupies least area in india **Histosols.**
366. Recently added / recently established **Gelisols.**
367. Highly weather soil in the world **Oxisols.**
368. Absent of Fe & Al oxides soil order is **Inceptisols.**
369. High N containing fertilizer **Anhydrous ammonia.**
370. High imported fertilizer in India **K.**
371. Fire hazardous fertilizer & explosive **Ammonium nitrate.**
372. Highest hygroscopic fertilizer **Ammonium nitrate.**
373. Equilent Basity of $NaNO_3$ **29.**
374. Equilent Basity of CaCN **63.**
375. Equilent acidity of NH_4Cl **128.**
376. Equilent acidity of $(NH_4)SO_4$ **110.**
377. Equilent acidity of urea **80-85.**
378. Equilent acidity of DAP **77.**
379. CaO liming material value is **179.**
380. $Ca(OH)_2$ liming material value is **136.**
381. Dominant state of red soil **Tamil Nadu.**
382. Dominant state of black soil **Maharashtra.**
383. Dominant state of saline soil **Gujarat.**
384. Dominant state of Alluvial soil **Uttar Pradesh.**
385. More surface area occupies in clay mineral **Vermiculite.**
386. High salt tolerant crops are **Barely, sugarbeet.**
387. Weathering stages **Initial > Juvenile > virile > senile > final.**
388. More than 8°C but less than 15°C and mean winter and summer temperature is >5° so it **Mesoic temperature regim.**
389. Which moisture regime having water throughout year **Udic.**
390. Weekly cemented by SiO_2 which endo pedon **Durinodes.**
391. Mixing of soil matrices with in the soil **Cryoturbation.**
392. Temperature always less than zero **Permafrost.**
393. Albic horizon material penetrate into agrillic **Tonguing.**
394. Dominant order in India **Inceptisol.**
395. Dominant Silicates in India **Phyllosillicates.**

396. Dominant moisture regime in India **Ustic.**
397. Dominant cation **Ca^{2+}.**
398. Dominant mineral in the earth crust **Feldspar.**
399. Dominant element in the earth crust **O_2** (Oxygen).
400. Nitrates in drinking water is not more than **10 Mg/L.**
401. Percentage of sulphur & N in ammonium sulphate **24% S, 20% N.**
402. Example for nitrification inhibiters **N- serve AM ST.**
403. Slow releasing N – fertilizers example **IBDU, CDU, Neem coated urea and Sulphur coated urea.**
404. Floculation order **$Al^{3+} > Ca^{2+} > H^{+} > Mg^{2+} > K^{+} > Mn$.**
405. Chelating power **Cu > Ni > Co > Zn > Fe > Mn.**
406. Ammonium sulphate suitable applied for which soils **Alkaline soils.**
407. Which one are cationic micronutrients? **Cu, Ni, Fe, Mn, Zn.**
408. Which one are Anionic micronutrients? **B, Mo, Cl.**
409. Examples for beneficial elements **Na, Co, Se, Va, Cr, Ti.**
410. Which elements are act as enzyme activator? **K, Ca, Mg, Mn, Zn.**
411. What is the N percentage in plants? **1-5% (10,000-50,000 ppm).**
412. Energy currency of the plants is which nutrient **P.**
413. Exact percentage of Cl in plants **100-500 mg.**
414. Cracks stem of celery deficiency symptoms due to **B.**
415. Frenching of tung grass due to **Mn.**
416. What is the percentage of N in mono ammonium phosphate? **11% N.**
417. Structural stability of protein is function of **S.**
418. Salt index ammonium sulphate **3.23.**
419. Soil crust strength measured by **Penetrometer.**
420. What is the hardness gypsum mineral? **2.**
421. What is the electrical conductivity of fresh water? **0.6.**
422. What is the electrical conductivity of Brakish? **1.5-3.**
423. What is the electron conductivity of Sea water? **45.**
424. Organic matter act as a source of **NPS.**
425. C:N ratio for young leaves / legumes **10:1 to 30:1.**
426. C:N ratio for Micro organism **5:1 to 10:1.**
427. C:N ratio for sawdust is **600:1.**

428. First successful commercial production **Superphosphate.**
429. First fertilizer plant came up **Ranipet (Chennai).**
430. N losses in FYM can be controlled to use preservatives like **Gypsum & SSP.**
431. What is the percentage of nutrient contains in FYM **0.5N : 0.2 P : 0.5 K.**
432. Phosphate solubility bacteria are **Bacillus, Pseudomonas.**
433. Phosphate solubility fungi is **Aspergillus.**
434. Potassium solubillizing bacteria **Bacillus mucilaginous.**
435. Celluloytic/lingolytic fungi is **Trichoderma viridae.**
436. Exfoliation also called as **Onion type of weathering.**
437. Weathering sequence proposed by **Goldich (1930).**
438. Weathering index proposed by **Jackson.**
439. Parent material is the initial stage of soil system at the time zero of soil formation given by **Jenny.**
440. Standard buffer solution pH are **4,7 & 9.2.**
441. Basic slag contains mostly **24 to 50% CaO.**
442. Press mud contains the pH of **6.5 -6.9.**
443. Press mud contains percentage of organic carbon**16-18%.**
444. Neutralizing value of press mud is **22-24%.**
445. Salt affected soil comes under which soil orders **Inceptisols, vertisols, mollisols, albisols.**
446. In Punjab state kallar soil are another name of which soils **Sodic soil.**
447. Gypsum equivalent value of pyrite **0.63.**
448. Gypsum equivalent value of $Cacl_2$ **0.85.**
449. Gypsum equivalent value of press mud **0.77.**
450. Press mud contains the percentage of **9% Ca, 24% S.**
451. In India the black soil is mostly fond in **Maharashtra.**
452. Red soil is mostly found in **Tamil Nadu.**
453. The soil which is most suitable for most crops is **sandy loam.**
454. Red colour in the red soils is due to presence of various oxides of **irons.**
455. The vertical section of the soil showing the various layers from the surface to the unaffected parent material is known as **soil profile.**
456. The arrangement of soil particles and their aggregate into certain defined is called **soil structure.**

457. Particle density of soil is also known as **true density.**
458. Generally the particles density of normal soil is **2.65 gram/cm³.**
459. The bulk density of normal soil is **1.33 g/cm³.**
460. The formula of porosity of soil is **porosity = 100-(bulk density / Particle density).**
461. The first soil classification system was developed by **Russian scientist Dokuchaiev (1880).**
462. The certain exchange capacity (CEC) of montmorillonite is **60-100 me / 100 gm.**
463. The U.S classification system the soil orders are **10.**
464. Formula of bulk density is **wt. of soil / volume of solids and pores.**
465. Law of minimum was given by **Von Liebig (1840).**
466. Top most mineral horizon is **A horizon.**
467. The term pH was introduced by **Sorensen (1909).**
468. Energy exchange elements are **H & O.**
469. K & Na is determined by **Flame Photometer.**
470. Micro nutrient is also known as - trace element, Oligo **Element or Spurned elements.**
471. Crop logging technique was given by **H. F. Clements.**
472. Metal nutrients are **K, Fe, Ca, Mg, Mn, Zn, Cu.**
473. Highly mobile nutrients are **N, P, K.**
474. Criteria of essentiality was proposed by **Arnon & Stout (1939).**
475. The term fictional nutrient was proposed by **Nicholas (1961).**
476. Moderate mobile nutrient is **Zn.**
477. FYM contains **0.5% N, 0.2 % P_2O_5 & 0.5 % K_2O.**
478. The P is available at **6-7 pH.**
479. Bio fertilizers are **Rhizobium, Azotobacter, Azospirillum , PSB, VAM, BGA, etc.**
480. In India the black soil is mostly fond in - **Maharashtra.**
481. Red soil is mostly found in – **Tamil Nadu.**
482. The soil which is most suitable for most crops is - **sandy loam.**
483. Red colour in the red soils is due to presence of various **oxides of iron.**
484. The vertical section of the soil showing the various layers from the surface to the unaffected parent material is known as **soil profile.**

485. The arrangement of soil particles and their aggregate into certain defined is called **soil structure.**
486. Particle density of soil is also known as **true density.**
487. Generally the particles density of normal soil is **2.65 gram/cm³.**
488. The bulk density of normal soil is **1.33 g/cm³.**
489. The formula of porosity of soil is - porosity **=100-(bulk density / Particle density).**
490. The certain exchange capacity (CEC) of montmorillonite is **60-100 me / 100 gm.**
491. The first soil classification system was developed by **Russian scientist Dokuchaev (1880).**
492. The U.S classification system the soil orders are **10.**
493. Formula of bulk density is = **wt. of soil / volume of solids and pores.**
494. Law of minimum was given by **Van Liebig (1840).**
495. Top most mineral horizon is **A horizon.**
496. Approx. 20 gram of rhizobium culture is required to treat **one kg seed.**
497. One packet of rhizobium culture contains **20 gram of rhizobium.**
498. Azotobacter is used for **Rice, cotton, sugarcane.**
499. Azospirillum bio fertilizer used for **Sorghum.**
500. The ability of soil to supply adequately the nutrients normally taken from the soil by plants **Soil fertility.**
501. The capacity of soil to produce crops **Soil productivity.**
502. 'Law of minimum' was given by **Justus von Liebig.**
503. Father of field plot technique **Jean Baptiste Boussingault.**
504. The criteria of essentiality of nutrients was given by **Arnon and stout.**
505. The definition of essential elements **D.J. Nicholas.**
506. The essentiality of nitrogen to plant given by **Theodore-Saussure.**
507. The essentiality of K, Ca, Mg and P to plant given by **C. Sprengel.**
508. The chemical compound or ion required by an organism **Nutrient.**
509. Beneficial elements examples: **Silicon, vanadium, cobalt and aluminium.**

510. Forms of nutrient elements absorbed by plants **Nitrogen Ammonium (NH_4^+) and Nitrate (NO_3^-), Phosphorus $H_2PO_4^-$ and HPO_4^{2-}, Boron H_3BO_3, $H_2BO_3^-$, HBO_3^{2-}, BO_3^{3-}, Molybdenum MoO_4^{2-}.**

511. The plants are normally composed of about 90–95% of **C, H, O.**

512. Micronutrient cations **Fe, Mn, Zn and Cu.**

513. Micronutrient anions **B, Mo and Cl.**

514. The plant nutrients which are mobile in plants **N, P, K and Mg.**

515. The plant nutrients which are partly mobile in plants **Fe, Zn, Mo and Cu.**

516. The plant nutrients which are immobile in plants - **Ca, S and B.**

517. The movement of plant nutrients along with flowing water - **Mass flow.**

518. The movement of plant nutrients due to concentration gradient - **Diffusion.**

519. 'Theory of contact exchange' was given by - **Jenny and Overstreet.**

520. CEC of plant roots - **10 to 100 cmol/ kg roots.**

521. Legumes have high root CEC and absorb more- **Divalent cations.**

522. Monocots have low root CEC and absorb more of - **Monovalent cations.**

523. The earth's atmosphere contain nitrogen about - **78%**.

524. The ploughed layer of majority of cultivated soils contains about - **0.02 to 0.04 % nitrogen.**

525. The N uptake is usually high in which form - **NO_3^-.**

526. The form of nitrogen which is less subjected to losses by leaching and denitrification - **NH_4^+-N.**

527. The conversion of organic nitrogen to mineral form or inorganic form – **Mineralization.**

528. The order of NH_4^+ fixation is as follows: **vermiculite > fine grained mica > smectites.**

529. The biological oxidation of NH_4^+ to nitrate is known as – **Nitrification.**

530. Conversion of NH_4^+ to nitrite is carried out by- **Nitrosomonas.**

531. The conversion from nitrite to nitrate is carried out by – **Nitrobacter.**

532. Nitrifying bacteria are- **Obligate autotrophic bacteria or chemo autotrophic bacteria.**

533. The form of nitrogen which is mostly subjected to losses by leaching and denitrification - **NO_3^-.**

534. Which bacterial growth is inhibited by the nitrification inhibitors to keep or retain the nitrogen in the NH_4^+ or ammonical form is – **Nitrosomonas.**

535. Nitrification inhibitors - **N – serve, DCD, AM and ATC.**

536. Natural nitrification inhibitors - **Neem cake and Karanj cake.**

537. Slow release N fertilizers - **Neem coated urea, Sulphur coated urea.**

538. The process of bacterial reduction of NO_3^- to NO, N_2O and N_2 gases is called as – **Denitrification.**

539. In assimilatory nitrate reduction, the nitrate reduced to - NH_3.

540. Denitrifying bacteria are–***Bacillus, Pseudomonas and Thiobacillus***.

541. The enzyme involved in biological nitrogen fixation - **Nitrogenase.**

542. The biological nitrogen fixation (BNF) contribute **- 140-170 MT N/yr.**

543. Symbiotic nitrogen fixer in legumes – **Rhizobium.**

544. Symbiotic nitrogen fixer in non legumes – **Frankia.**

545. Availability of phosphorus is optimum in the soil pH range - **6.1 to 7.0.**

546. The NH_4^+ fixation will be restricted by the presence of - **K^+.**

547. Azolla – Anabaena symbiosis can fix N_2 upto - **30-40 kg N ha^{-1}.**

548. *Rhizobium* sp. involved in nodulation of soybean - ***R. japonicum***.

549. *Rhizobium* sp. involved in nodulation of Alfalfa - ***R. meliloti***.

550. Losses of nitrogen through - **Leaching, volatilization, Denitrification and Immobilisation.**

551. NH_4^+ or K^+ fixation is more in which type of clay mineral **– 2: 1 clay minerals.**

552. N is an essential component of **- Amino acids, proteins, nucleic acids and alkaloids.**

553. Inverted 'v' shape symptom is due to deficiency of **– N deficiency**.

554. Most widely used source of hydrogen in India used for preparation of nitrogenous fertilizers is – **Naptha.**

555. The concentration of urea in foliar application is advocated in dry land areas. **- 2%**

556. The first organic fertilizer – **Urea.**

557. The maximum permissible limit of biuret in the urea fertilizer - **1.5%**.
558. The nitrogen content in Ammonium sulphate - **21.0 %**.
559. The nitrogen content in CAN – **25%**.
560. The nitrogen content in Diammonium phosphate (DAP) – **18%**.
561. The chemical formulae of Flourapatite is - **$3\ Ca_3\ (PO_4)_2\ CaF_2$**.
562. The chemical formulae of Monocalcium phosphate is - **$Ca\ (H_2PO_4)_2$**.
563. The chemical formulae of Tricalcium phosphate is - **$3\ Ca_3\ (PO_4)_2$**.
564. Organic form of phosphorus are **Inositol phosphate, Phospholipids and Nucleic acids etc.**
565. The $H_2PO_4^-$ ions dominate in soils at soil pH - **<7.0.**
566. The PO_4^{3-} ions dominate in soils at soil pH- **>7.0.**
567. At what soil pH, the concentration of $H_2PO_4^-$ is equal to that of HPO_4^{2-} - **7.2.**
568. Optimum soil pH of for availability of phosphorus - **6.5 to 7.5.**
569. P fixation demonstrated by - **Thomas Way.**
570. In acid soils (pH < 6.5) P fixation due to - **Iron and Aluminium.**
571. In alkaline soils (pH < 6.5) P fixation due to – **Calcium.**
572. P is fixation is more in in the crystal lattices of - **1:1 type of clay minerals (Kaolinite).**
573. Quantity– Intensity relationships of P was given by – **Schofield.**
574. The element which is essential constituent of number of enzymes and important in energy transfer – **P.**
575. The element which increases rhizobial activity, increases the formation of root nodules thus helping in more N – fixation - **P.**
576. Purple colour discolouration of plants due to **– P deficiency.**
577. Excess of P causes deficiency of - **Fe and Zn.**
578. Phosphorus is absorbed by plant from soil mostly by the process - **Diffusion.**
579. The chemical formula for potassium metaphosphate is - **KPO_3**.
580. The P content in DAP – **46%**.
581. The chemical formulae of orthoclase - **$KAl\ Si_3O_8$**.
582. The chemical formulae of muscovite - **$KAl_3\ Si_3O_{10}\ (OH)_2$**.
583. The chemical formulae of biotite - **$K(Mg\ Fe)_3\ Al\ Si_3\ O_{10}\ (OH)_2$**.

584. The percentage contribution of Slowly available K in total K - **1-10% total K.**
585. 'Lattice Hole' theory was proposed by - **Page and Baver.**
586. The concept of quantity, intensity of K proposed by- **Beckette.**
587. A measure of K in soil solution that is immediately available for absorption by plants - **Intensity factor.**
588. The crops tend to absorb and accumulate potassium for in excess of their needs is called as- **Luxury consumption.**
589. Muriate of potash is produced mainly from the ore containing - **Sylvinite.**
590. The mineral that contain K_2SO_4 - **Langbeinite.**
591. K functions - **opening and closing of stomata, drought tolerance, enhances crop quality, shelf life of fruits and vegetables, Imparts disease resistance.**
592. Burning appearance of leaves due to deficiency of - **K.**
593. Muriate of potash contains - **60% K_2O.**
594. Sulphate of potash contains - **50% K_2O.**
595. Muriate of potash cannot be applied to crops - **Tobacco, grapes, cotton, sugarcane, potatoes, tomatoes.**
596. The nutrient contents in Ammonium sulphate - **20 % N and 24% S.**
597. Single super phosphate contain - **16 % P_2O_5, 21 % Ca and 12% S.**
598. Potassium sulphate contain- **52% K_2O and 18% S.**
599. The calcium concentration of the earth's crust is about - **3.20%.**
600. Calcium containing primary minerals - **Augite, Hornblende , Anorthite, Epidote.**
601. The dominant cation in normal cultivated soils - **Calcium.**
602. The chemical formulae of Gypsum - **$CaSO_4.2H_2O$.**
603. The chemical formulae of Apatite - **$Ca_{10}(PO4)_6$.**
604. Legumes use much more calcium than grasses due to - **high CEC.**
605. The base saturation required by Montmorillonite to release Ca **-70%.**
606. The base saturation required by kaolinite to release Ca - **40%.**
607. Blossom end rot disorder in tomatoes due to deficiency of **- Calcium.**
608. Bitter pit in apple due to deficiency of - **Calcium.**

609. The nutrient essential for the formation of cell wall and meristematic activity - **Calcium.**
610. The magnesium concentration of the earth's crust is about - **1.93%**.
611. Magnesium containing primary minerals - **Biotite, dolomite, hornblende, olivine, epsomite ($MgSO_4$. 7 H_2O) and serpentine.**
612. Ideally, Ca / Mg ratio should not be greater than - **7:1.**
613. High levels of exchangeable K^+ can interfere with uptake of – **Mg.**
614. For field crops K / Mg ratio should be - **< 5:1.**
615. Element is bonded in the centre of each porphyrin ring in the chlorophyll - **Mg.**
616. About 70 % of Mg is associated with anions such as - **Malate and citrate.**
617. The nutrient essential for uptake and translocation of phosphorus and movement of sugars within the plants - **Mg.**
618. Cattle consuming forages with low Mg may suffer from "Hypomagnesemia" commonly known as - **Grass tetany.**
619. The earth's crust contains sulphur about - **0.06%**.
620. More than 95% of total sulphur in soil is present as - **Organic Sulphur.**
621. SO_4^{2-} ions are more prone for leaching when associated with - **Monovalent cations (Na^+).**
622. Mineralisation of sulphur takes place at C/S ratio - **< 200 : 1.**
623. Immobilisation of sulphur takes place at C/S ratio - **> 400: 1**.
624. Chemolithotrophic sulfur oxidizing bacteria - ***Thiobacillus thiooxidans. Thiobacillus thioparus*.**
625. Sulphur reduction or Sulphide injury occurs at a redox potential of - **< -150 mV**.
626. H_2S toxicity injury to rice commonly called as - **Akiochi.**
627. S containing amino acids are - **Cystein and methionine.**
628. N:S ratio of plants is between - **9 to 12 : 1.**
629. Reduced synthesis of proteins and oil in plants is due to deficiency of - **S.**
630. The micronutrients essential for green plants are- **Iron, copper, manganese, boron, zinc, molybdenum, chlorine and nickel.**
631. The total essential nutrients for plant growth - **17.**

632. The seventeenth essential element included in the list of nutrient element - **Nickel.**

633. The element have a beneficial effect on rice plants - **Si.**

634. Zinc containg minerals - **Sphalerite (ZnS), Smithsonite $ZnCO_3$.**

635. Copper containg minerals **Chalcocite (Cu_2S), Azurite {$Cu_3(OH)_2(CO_3)_2$}.**

636. Iron containg minerals - **Haematite (Fe_2O_3), Goethite (FeOOH), Magnetite (Fe_3O_4), Pyrite (FeS_2).**

637. Manganese containg minerals – **Pyrolusite(MnO_2), Manganite (MnO OH).**

638. Boron containg minerals - **Borax ($Na_2B_4O_7 10H_2O$), Tourmaline {$Na(MgFe)_3.Al_6(BO_3)_3Si_6O_{18}(OH)$}.**

639. Molybdenum containg minerals - **Molybdenite (MoS_2), Wulfenite ($PbMoO_4$).**

640. Micronutrients pools of in soils postulated by - **Viets.**

641. The pool contains water soluble ions - **Pool A.**

642. The pool contains exchangeable micronutrient cations - **Pool B.**

643. The pool includes the adsorbed, chelated or complexed ions - **Pool C.**

644. The pool includes micronutrient cations in secondary clay minerals and insoluble metal oxide - **Pool D.**

645. The pool Comprises cations held in primary minerals - **Pool E.**

646. The element occur in all the pools - **Mo.**

647. Availability of molybdenum is more in - **Alkaline pH.**

648. Availability of Micronutrients **(Iron, copper, manganese and zinc)** is more in - **Acidic pH.**

649. Over liming of acid soils causes deficiency of micronutrient cations like - **Zn, Fe, Mn and Cu.**

650. Water logged condition reduce the availability of **Zn and Cu** and increase availability of - **Fe and Mn.**

651. Average zinc content of mineral soils varies between - **10 and 330 ppm.**

652. Zn deficiency cause - **Khaira disease of rice, White bud of maize, Mottle leaf or Frenching of citrus and little leaf of brinjal and mango.**

653. High Zn susceptibility varities of rice - **Jaya, IR - 8, RP-4.**

654. The Critical level of 0.005M DTPA extractable zinc - **0.5 to 1.0 ppm.**
655. Copper concentration in earth's crust ranges from - **55 to 70 ppm.**
656. The nutrient deficient in soils having higher organic matter - **Copper.**
657. The Critical level of 0.005M DTPA extractable Copper - **0.2 to 0.5 ppm.**
658. The element enhances the fertility of male flowers - **Copper.**
659. Exanthema and dieback in citrus due to deficiency of - **Copper.**
660. Empty glumes in wheat and reclamation disease is due to - **Copper deficiency.**
661. Bordeaux mixture contain essential element - **Copper.**
662. Fourth most abundant element in the lithosphere and comprises of about 5 % of the earth's crust - **Iron.**
663. The Critical level of 0.005M DTPA extractable iron - 2.5 to 5.8 ppm.
664. For each unit increase in pH, the activities of Fe^{3+} and Fe^{2+} decrease by - **1000 fold and 100 fold respectively.**
665. Flooding and submergence of soils improve availability of - **Iron.**
666. Iron is a constituent of - **leg haemoglobin, Ferredoxin and peroxidise.**
667. Interveinal chlorosis appearing first on the younger leaves with leaf margins and veins remaining green is due to deficicency of - **Iron.**
668. Fe – EDDHA (Ethylene Diamine Dihydroxy acetic acid salt of Fe) use for - calcareous soils.
669. Iron chelate suitable for reclamation of iron in acid soil - **Fe - EDTA (Ethylene Diamine Tetra Acetate).**
670. Iron chelate suitable for reclamation of iron in alkaline soil - **Fe-DTPA (Diethylene Triamine Penta Acetate).**
671. Iron chelate suitable for reclamation of iron in alkaline soil - **Fe – HEDTA (Hydroxy Ethylene Diamine Tetra Acetate).**
672. Leaf bronzing in rice is due to - **Fe toxicity.**
673. Manganese concentration in the earth's crust averages - **1000 ppm.**
674. Soil submergence and water logging lower the redox potential and increase the amount of - **Soluble Mn^{2+}.**
675. The element associated with photosystem II and TCA cycle - **Mn.**

676. Critical limit of 0.005M DTPA extractable Mn in soil - **2-4 ppm.**
677. The boron content in Indian soils ranges - **7 to 630 ppm.**
678. Crops most sensitive for boron deficiency are - **sugar beet and celery.**
679. The element plays a role in pollen germination and pollen tube growth - **Boron.**
680. Boron deficiency causes - **Heart rot of sugar beet.**
681. Boric acid H_3BO_3 contains B - **17%.**
682. Borax ($Na_2 B_4 O_7 10 H_2O$) contains B **-10.6%.**
683. The total Mo content of most soils varies from - **0.2 to 5.0 ppm.**
684. The available Mo content of soils extractable with - **Ammonium oxalate (Tamm's extractant) at pH - 3.3.**
685. Critical limit of Ammonium oxalate extractable molybdenum in soil - **0.05 to 0.2 ppm.**
686. Liming of acid soils increases availability of **– Mo.**
687. The element which is a constituent of nitrate reductase and helps in enzymatic transfer of e- for reduction of N **– Mo.**
688. Whiptail in Cauliflower due to deficiency of nutrient **– Mo.**
689. Plants which are sensitive to chloride **- Tobacco and tomatoes *etc*.**
690. The consumption of water containing high levels of nitrate N leads to 'methaemoglobinemia' called as **-Blue baby syndrome.**
691. The World Health Organisation standard for drinking water is - **10 mg NO_3-N/L.**
692. The process of enrichment of surface water bodies with nutrients - **Eutrophication.**
693. Eutrophication is due to nutrients - **Both N and P.**
694. Eutrophication is mainly due to – **phosphorous.**
695. Rock phosphate being highly potential source for heavy metals - **lead and cadmium.**
696. 'Law of Restitution' and law of minimum was proposed by **- Justus Von Liebig.**
697. The most critical stage of growth for tissue testing is - **at the time of bloom or from bloom to early fruiting stage.**
698. The rapid tissue test for the nitrates estimation in plant - **Diphenylamine test.**

699. The rapid tissue test for the Phosphates estimation in plant - **Molybdate + Stannous oxalate test.**

700. The rapid tissue test for the potassium estimation in plant - **Sodium cobalti nitrate test.**

701. The range of nutrient concentration at a specified growth stage above which the crop is amply supplied and below which the crop is deficient - **Critical Nutrient Concentration.**

702. DRIS - **Diagnosis and Recommendation Integrated System.**

703. DRIS proposed by – **Beaufils.**

704. The soil fertility evaluation technique not only the limiting nutrient but also the order in which the nutrients are likely to become limiting – **DRIS.**

705. In DRIS, Horizontal arrows (→) in the inner circle indicate- **balance between nutrients.**

706. In DRIS, Vertical (→) arrows representing - **nutrient imbalance.**

707. Certain plants are very sensitive to deficiency of a specific plant nutrient and they produce specific symptoms are called as - **Indicator plants.**

Element	Indicator plant
N	Cauliflower, Cabbage
P	Rape seed
K	Potato
Ca	Cauliflower, Cabbage
Mg	Potato
Fe	Cauliflower, Cabbage, Potato
Na	Sugar beet
Mn	Sugarbeet, Oats, Potato
B	Sunflower
Zn	Maize
Cu	Wheat

708. In biological methods of soil fertility evaluation, the **Sackett and Stewart** technique used to find out available nutrient status of - **P_2O_5 and K_2O.**

709. Mehlich's cunninghamella plaque method used to find out available nutrient status of - **P_2O_5.**

710. Mulder's *aspergillus niger* test for estimation of - **Cu and Mg.**

711. In Mehlich technique for determining available K_2O content, the microorganism used is - ***Aspergillus niger***.

712. The test crop for jenny pot culture technique – **lettuce.**

713. Soil plaque method was proposed by - **Sackett and Stewart technique.**

714. The amount of nutrients in soil which behave in a similar way as the applied fertilizer nutrient doses - **A value.**

715. **Radioactive isotopes** technique of A value calculated by the formula.

716. A= B(1 – Y)/Y.

717. A-value was proposed by - **Fried and Dean.**

718. The techniques is used for assessing available P in soil and does not include growing plants - **E value.**

719. Soil test based fertilizer recommendations or Fertility gradient approach was given by – **Ramamoorthy.**

720. The critical soil test level concept given by - **Cate and Nelson.**

721. Empirical equations for scheduling fertiliser P dosage to crops was given by - **Krishnamoorthy *et. al*.**

722. According to Krishnamoorthy empirical equations, the percentage extractable P in black cotton soil = **100 – Clay percentage.**

723. According to Krishnamoorthy empirical equations, the percentage extractable P in red and alluvial soils soil = **100 – 2 (Clay percentage).**

724. Yield per unit input is defined as - **Nutrient use efficiency (NUE)**

725. The word "Manure" is originated from the French word - **"Manoeuvrer".**

726. The substances which are organic in nature, capable of supplying plant nutrients in available form , bulky in nature and having no definite composition is called – **Manure.**

727. A mined or manufactured material containing one or more essential plant nutrients in potentially available forms in commercially valuable amounts is called – **Fertilizer.**

728. The word "Fertilizer" was derived from Latin word - **"Fertil".**

729. Examples of Bulky organic manures – **Manures, Composts, Green manures, Agro-based Industry wastes and Municipal wastes.**

730. Examples of Concentrated organic manures - **oil cakes, Blood meal , Bone meal, Horn meal, Guano.**

731. Edible oil cakes are - **Mustard cake, Sesame cake, Linseed cake.**

732. Non –edible oil cakes are - **Castor cake, Neem cake, Sunflower cake, Mahua cake, Karanj cake.**

733. Concentrated organic manures contain higher - **Nitrogen content.**

734. Concentrated organic manures (oil cake) contain higher amount of nitrogen - **Safflower cake.**

735. Adult old cattle provide better manure than – **young one.**

736. Trench method of preparing FYM was recommended by - **C. N. Acharya.**

737. Maximum losses of nutrients occur in storage method – **Heap method.**

738. Chemical preservatives are added to FYM to decrease nitrogen losses - **Gypsum and Super phosphate.**

739. The word compost is derived from the Latin word **"COMPONERE".**

740. A process of converting organic matter in to manure in a short time by accelerating fermentation process under controlled conditions is called – **Composting.**

741. A typical compost earthworm is - ***Eisenia foetida***.

742. ADCO - **Agricultural Development Company**.

743. Agricultural Development Company was initiated and developed by - **H.B. Hutchinson and E.H. Richards.**

744. Starters in ADCO process are - **Ammonium sulphate, Ammonium phosphate, Super phosphate, Muriate of potash, Ground limestone, urea.**

745. The compost produced in ADCO process is called as- **synthetic FYM.**

746. Activated compost method was developed by - **Fowler and Ridge.**

747. Indore process developed in India by - **Howard and Ward.**

748. Bangalore process of composting was developed by - **Dr. C.N. Acharya.**

749. The composting process which involves Aerobic and anaerobic fermentation process- **Bangalore process.**

750. Starter material in rural compost - **Dung and water.**

751. Starter material in urban compost - **night soil.**

752. The chemical used in urban compost to prevent the offensive smell - copper sulphate.
753. Scientific method of breeding and rising earthworms in controlled conditions is – **Vermiculture.**
754. The crops which are grown and buried in the same field which is to be green-manured, either as a pure crop or as intercrop with the main crop - **Green manure crops.**
755. Green manure crops - **Sunn hemp, Daincha, Wild indigo, Indigo, Pillipesara.**
756. The green manure crop which is suitable for reclamation of alkali or sodic soils – **Daincha and *Sesbania speicosa*.**
757. Green manure pulse crop - **Pillipesara.**
758. Green manure crop suitable for poor and hard soils- **Horse gram.**
759. Keeping the cattle and sheep in the fallow land after the harvest of the last crop, throughout day /and or night provided with suitable food and shelter is called – **Penning.**
760. **Poudrette** - is the product obtained from night soil without any admixture of other organic waste materials.
761. The water consists of liquid and solid waste including night soil – **Sewage.**
762. The clear supernatant liquid obtained after aeration during sedimentation process is – **Effluent.**
763. The sediment that settles down in the activated sludge process is - **Sludge.**
764. The water drained from the kitchens, bathrooms and drainage water of the Streets - **Sullage.**
765. Best method of storage of FYM – **Covered pit method.**
766. The distance between gas plant and place where gas is to be used should be with in- **20 meters.**
767. The major constituent of bio gas is - **CH_4.**
768. Gobar gas production was first started in – **Germany.**
769. The first person attempt to produce gobar gas in India - **S.V. Desai.**
770. Green Leaf Manure crops - **Glyricidia, Subabul, Cassia.**
771. Bone meal is effective in – **Acidic soils.**
772. Nitrogen content in horn meal - **14%.**
773. The process of removal of proteins from the grist – **Deamination.**

774. Nitrogen content in blood meal- **10-14%**.
775. Meat meal also referred as – **Tankage.**
776. A product of sea bird (Pelican, Gannets, and Albatrosses) excrement is – **Guano.**
777. Fertilizer which contains only one primary or major nutrient - **Straight fertilizers.**
778. Deposits of natural nitrates present in vast areas in **CHILE and PERU of pacific.**
779. Coast are called **- chile salt peter.**
780. Ammonia synthesis by **- Haber –Bosch process.**
781. Nitrate synthesis by **- arc-gap process.**
782. The energy needed to get 1 kg of 'N' in the form of fertilizer - **40,000 kilo joules.**
783. The temperature required to maintain in "Haber – Bosch" process **- 550°C temp.**
784. The pressure required in "Haber – Bosch" process - **200 atm.**
785. The catalyst used in ammonia synthesis by Haber – Bosch" process **- Activated iron oxide.**
786. The plant used coal as a source H_2 **- 'Ramagundam'.**
787. The plant used water as a source of H_2 – **NFL, Nangal (Punjab).**
788. The light distillation fraction of petrol (or) crude oil with boiling point 215°C - **Naptha.**
789. The main source of hydrogen in India – **Naphtha.**
790. About 64 % of ammonia produced in the world is from **- natural gas.**
791. Natural gas is a mixture of gases- **methane , ethane , butane and CO_2** etc.
792. Cheapest source of N fertilizer – **NH_3**.
793. Ammonia is soluble in water up to 30% by weight and containing 24 per cent nitrogen called **- Aqua ammonia.**
794. The oldest synthetic N fertilizer **- Ammonium sulphate.**
795. The fertilizer produced as by product of steel industry **- Ammonium sulphate.**
796. The amount of $CaCO_3$ is needed to neutralize the acidity created by 100 kg of $(NH4)_2\ SO_4$ **- 110 kg.**

797. Calcium ammonium nitrate also called as - **Lime Ammonium Nitrate.**
798. Neutral fertilizer – **CAN.**
799. CAN is a mixture of - **NH_4NO_3 and $CaCO_3$.**
800. CAN contains **25 %** nitrogen with equal quantity of each NH_4^+ and NO_3^- forms of nitrogen - **12.5% and 12.5%.**
801. Explosive fertilizer is - **Ammoniun nitrate (33% N).**
802. During the manufacture of CAN, explosive and hygroscopic nature of ammonium nitrate reduced by addition - **lime powder.**
803. The intermediate and unstable compound produced in urea manufacture - **Ammonium carbamate.**
804. Equivalent acidity of urea - **80 kg of $CaCO_3$.**
805. The fertilizer which is considered as organic fertilizer – **Urea.**
806. 82% of N requirement of the crops in India is supplied by - **urea.**
807. When temperature exceeds 1000C, urea produces a toxic product called - **biuret.**
808. The amount of biuret content in urea should not exceed - **1.5%.**
809. Oldest N fertilizer, best for top dressing in rice - **Ammonium sulphate (20.6% N and 24% S).**
810. Most concentrated nitrogenous fertilizer - **Anhydrous ammonia (81% N).**
811. The fertilizer which destroys soil aggregates - **Sodium nitrate.**
812. Paddy takes N in the form of - **NH_4^+.**
813. Fertilizer application in lowland paddy is done at - **Reduced zone.**
814. In India phosphate rock deposits have been found in - **Udaipur in Rajasthan and Mussoorie in Uttaranchal.**
815. The basic raw material for manufacturing any phosphatic fertilizer -**Rock phosphate.**
816. Water soluble phosphatic fertilizers - **SSP, DSP, TSP and DAP.**
817. Citric acid soluble phosphatic fertilizers - **DCP, Basic slag and Bone meal.**
818. Water and citrate insoluble phosphorus fertilizers - **Rock phosphate, Raw bone meal.**
819. In water soluble P fertilizers, P is present in the form of - **Monocalcium phosphate [$Ca(H_2PO_4)_2$].**

820. SSP is suitable for - **neutral to alkaline soils.**
821. Basic slag and Bone meal are suitable for - **acidic soils.**
822. Rock phosphate is well suited for - **highly acidic soils.**
823. The mineral in phosphate rock is - **apatite [$Ca_{10}(PO_4)_6$].**
824. The oldest artificially produced fertilizer - **Single super phosphate.**
825. SSP has - **2/5 MCP and 3/5 Gypsum by weight.**
826. SSP also contains - **16%P , 21.0% calcium, 12% sulphur.**
827. SSP = **Rock phosphate + H_2SO_4.**
828. TSP = **Rock phosphate + H_3PO_4.**
829. Nitro phosphates = **Rock Phosphate +HNO_3+NH_3.**
830. There are 2 grades of SSP available in India- **14% P_2O_5 and 16% P_2O_5.**
831. Basic slag is a by product of steel industry (in open hearth process) is called - **Thomas slag.**
832. The mineral is added at the rate of 3.4 kg per tonne of lime to reduce viscosity of the Basic slag – **Fluorospar.**
833. Basic slag enriched with rock phosphate and marketed as **– pelophos.**
834. Element pottasium with symbol K means - **Kalium.**
835. **Chemical formulae of K minerals.**
836. Carnalite – **KCl $MgCl_2.6H_2O$.**
837. Schoenite – **$K_2SO_4.MgSO_4$.**
838. Sylvite – **KCl.**
839. Sylvinite – **KCl.NaCl.**
840. Langbeinite – **$K_2SO_4.2MgSO_4$.**
841. Double salt of $K_{2S}O_4$ and Mg SO_4 is – **Schoenite.**
842. Geological Survey of India has reported the occurrence of K-deposits in - **Leh (Jammu and Kashmir) and Bikanir (Rajasthan).**
843. The principle raw material Manufacturing of MOP **- Sylvinite.**
844. Potassic fertilizer containing highest amount of K_2O- **Muriate of potash (60% K_2O).**
845. Cheapest K fertilizer – **MOP.**
846. The K fertilizer not suitable for tobacco and potato **– MOP.**
847. The term muriate is derived from - **Muriatic acid.**

848. Removal of clay particle form sylvinite is called - **Desliming.**
849. Potassic fertilizer suitable for fertigation - **Potassium Nitrate (44% K_2O).**
850. The principle raw material Manufacturing of SOP- **Langbeinite (K_2SO_4 2Mg SO_4).**
851. The by product evolved in Mannheim furnace process - **HCl.**
852. The most imported fertilizer - **K fertilizers.**
853. Straight fertilizer - **Urea/Potassium Chloride/Ammonium Sulphate.**
854. Complex fertilizer - **Diammonium phosphate/Ammonium phosphate.**
855. Neutralizing value of Quick lime is - **179.**
856. Neutralizing value of Slaked lime is - **136.**
857. Neutralizing value of Calcium carbonate is - **100.**
858. Neutralizing value of Dolomite is - **109.**
859. The application of fertilizer along with irrigation water - **Fertigation.**
860. Nitrification inhibitors - **Oxamide (31% N) and Thiourea.**
861. Slow released N fertilizers - **S coated urea, Neem coated urea, Urea super granule.**
862. Rhizobium fixes atmospheric nitrogen/ha to the soil - **50-100 kg N/ha.**
863. The application of the same fertilizer more than once to a crop is known as - **Split application.**
864. The fertilizers containing two or more major plant nutrients, which are in chemical combination are called as - **complex fertilizers.**
865. The fertilizers contain only two major nutrients such as N and P_2O_5 in chemical combinations is known as - **Incomplete complex fertilizer.**
866. Those fertilizers having all the three major nutrients viz., N, P and K - **Complete fertilizers.**
867. Fertilizers contain more than 25% of primary nutrients - **High analysis fertilizers.**
868. Mono ammonium phosphate [MAP] = **NH_3 + H_3PO_4 [1:1 ratio].**
869. Diammonium phosphate [DAP] = **NH_3 + H_3PO_4 [2:1].**
870. Ammonium phosphate sulphate [APS] = **H_3PO_4 +H_2SO_4 + NH_3.**

871. Ammonium poly phosphate [APP] = **Ammonia + Super phosphoric acid.**

872. Urea ammonium phosphate [UAP] = **Phosphoric acid + Ammonia + Urea.**

873. Urea ammonium phosphate is popularly called as **- Uramphos.**

874. Nitro phosphates = **HNO_3 + Phosphate rock + Ammonia.**

875. All the properties of fertilizer which affect the flow of fertilizers through distribution equipment are collectively called – **Drillability.**

876. Cracking in fertilizer mixture can be reduced by adding - **conditioners.**

877. A process of formation of new compounds due to crystallization with a change in temperature and moisture of the fertilizer when the fertilizer is under pressure is called **- Crytal knitting.**

878. The ratio of the increase in osmatic pressure produced by the given fertilizer to that produced by the same weight of $NaNO_3$ is - **Salt index.**

879. The part of fertilizer available to succeeding crop other than for which it is applied is called **- residual effect.**

880. The physical mixture of solid fertilizers materials containing 2 or more major plant nutrients are called as **- Mixed fertilizer or bulk blended fertilizers (BBF).**

881. The reaction occurs when two compounds without a common ion are mixed in the presence of moisture - **Double decomposition.**

882. The materials added to non-granular or granular mixtures to improve their physical condition and to decrease their caking – **Conditioners.**

883. Examples of filler materials **- Sand, Saw dust, Dolomite, Soil etc.**

884. Herbicide –fertilizer mixture commonly called as **- weed and feed mixture.**

885. The minimum guaranteed quantity of plant nutrient content in terms of N, available P_2O_5 and available K_2O present in a given multinutrient fertilizer is known as- **Fertilizer grade.**

886. The relative percentage of N_2, P_2O_5 and K_2O in a fertilizer - **Fertilizer ratio.**

887. The cost per unit of the plant nutrient contain in a given fertilizer - **unit value of fertilizers.**

888. The cheapest fertilizer based on unit value- **Urea.**
889. Calcium nick named as - **Root developer.**
890. The mineral supplying magnesium - **Dolamite ($CaCO_3$ $MgCO_3$) - 12%.**
891. High acidic nature of the soil induces - **calcium and magnesium deficiency.**
892. Excess P leads to deficiencies of - **Zn, Fe, Cu.**
893. More the sulphate present in the soils, lesser the availability of – **Mo.**
894. Over liming causes lesser the availability of - **Fe, Cu, Mn, Zn.**
895. Copper sulphate is also called - **Blue Vitriol.**
896. The metallic molecules of varying sizes and shapes in which the organic part binds the nutrient in a ring like structure - **Chelates.**
897. EDTA :**Ethylene Diamine Tetra Acetic Acid.**
898. HEDTA - **Hydroxy Ethylene Diamine Tetra Acetic Acid.**
899. EDDHA - **Ethylene Diamine Dihyroxy Acetic Acid .**
900. **DTPA - Diethylene Triamine Penta Acetic Acid.**
901. Fertilizer control order passed in year – **1957.**
902. The symbiotic association of Azolla pinnata and Anabaena azollae is termed as - **Azolla Anabaena complex.**
903. The biofertilizer used for irrigated rice crop - **Blue green algae or Cyanobacteria.**
904. Nitrogen fixation by BGA have been estimated to be - **25-30 kg N/ha.**
905. VAM (vesicular –arbuscular –mycorrhiza) is - **obligate symbiosis.**
906. VAM genera- **Glomus, Gigaspora, Enterospora etc.**
907. VAM fungi enhancing the nutrient uptake of - **P, Zn and S.**
908. Phosphorous solubulizing fungi – ***Pencillium and Aspergillus***.
909. Phosphorous solubulizing bacteria – ***Bacillus and Pseudomonas***.

Manufacturing Processes of Fertilizers

FERTILIZER	MANUFACTURING PROCESSES
Ammonia	Haber –Bosch process
Nitrate	Arc-gap process
Ammonium Sulphate	By product process, Neutralization process, Gypsum process or leuna process

Urea	Wholer process or carbamate strippers process.
Single Super Phosphate	Den process, Continuous rock phosphate acidulation process
Triple super phosphate	Cone mixer process
Basic Slag	Open - hearth process
MOP	Crystallization process, Flotation process
SOP	Langbeinite process, Mannheim furnace process
Nitro Phosphates	ODDA process, PEC process.

1. **Tolerant to acidic soils** – Wet rice, potato, mustard.
2. **Tolerant to saline soils** – Chillies, cucurbits, wheat, sorghum, bajra, cluster beans, barley etc.
3. **Tolerant to alkali/sodic soils** – Barley, cotton, Bengal gram, berseem, sunflower, maize, etc.
4. **Tolerant to waterlogged soils** – Wet rice, daincha, para grass, napier grass, guinea grass.
5. **Crops tolerant to soil erosion** – Marvel grass, groundnut, black gram, rice bean, moth bean, and horse gram.
6. **India** at present is the second largest producer of fertilizer – nitrogen in the world, and enjoys the third position for phosphate fertilizers. However, potash is totally imported.
7. The consumption of chemical fertilizers (in terms of nutrients) during 2010-11 has been **282 lakh tones** comprising of **166 lakh tonnes** of nitrogen, **81 lakh** tonnes of phosphatic and **35 lakh tones** of potassic fertilizer.
8. **The all – India average consumption** of fertilizers has increased from 95 kg per ha in 2004-05 to 144 kg per ha in 2010-11. While per hectare consumption is 237.1 kg in Punjab and 225.7 kg in Andhra Pradesh, it is comparatively low in MP (81 kg/ha), Orissa (58 kg/ha), Rajasthan (48.3 kg/ha) and Himachal Pradesh (54.8 kg/ha) and below 5 kg/ha in some of the North Eastern States.
9. Apni Mitti Pahachane Abhiyan of **Uttar Pradesh**.
10. National Project on Management of Soil Health & Fertility.
11. BHOOCHETNA - **Maize Intercropping with Red Gram – Karnataka.**
12. Fire explosive fertilizer is - **Ammonium nitrate.**
13. An organic fertilizer - **Urea.**

14. Responsible for the hardness of water- **Ca and Mg.**
15. P is solubilized by the production of **organic acids like citric, fumaric, malic** and other acids while organic P is mineralized through production of enzymes like acid and **alkaline phosphatases.**
16. **CEC** of illite- 10 meq/100 g, Kaolinite-3-10, Montmorrillonite-80-100, Vermicullite-80-120.
17. **Mycorhiza** is a symbiotic association between – Fungi with root system of higher fungi belonging to Endogenaceae (***Glomusetunicatum, G. agrocarpus***) family existing as **obligate root symbionts,** infect and develop vesicles and arbuscules inside the roots of wide range of plants except members of **Cruciferaceae and Chenapodeacea.**
18. **Sand** : at least 70% sand and < 15% clay, **Silt** : at least 80% silt and < 12% clay, **Clay**: at least 35% clay.
19. **Opal** is mostly of plant origin and called biogenic silica.
20. **Cationic nutrients** (NH_4^+, Ca^{2+} and K^+ etc) are held/ stored by negatively charged soil clay colloids **unlike anionic nutrients** (NO^{3-}, $S)_4^{2-}$ and Cl^-) which are prone to leaching losses under flooded / high rainfall conditions.
21. Approx. one million worms in an acre produce - **500 kg/ day/ acre – 200t / acre/ year.**
22. **S** toxicity causes Akiochi disease in Rice. **H_2S, being a respiratory poison,** is the principal compound formed in sustained submerges soils.
23. P deficient plants accumulate **anthocyanin pigment.**
24. **Zn is an activator for several enzyms** like alcoholic dehydrogenase- ADH, glutamic dehydrogenase – GDH, carbonic anhydrase, RNA and DNA polymerases, Cu- Zn superoxide dismutase.
25. **Mn is a constituent of plant compound called manganin.** It plays a crucial role in water splitting process (Hill reaction) and consequent O_2 evolution (photosystem II).
26. **Alluvial soils** are rich in potash and calcium but are deficient in organic matter.
27. **Green Manuring (GM) is practised** in 6 m/ ha as against potential of 25 m ha.

28. In 2010-11, there exist **1049 soil testing laboratories** in the country with a soil analysing capacity of 106 lakh soil samples per annum.
29. S is a constituent of **cysteine, cysteine, methionine.**
30. **Methionine** is a principal precursor in biosynthesis of ethylene and **phytosiderophores which is a chelator of Fe, Mn and Zn.**
31. Green manuring is good source of **N (2-3%) with low C: N ratio of about 20.**
32. S containing vitamins like **biotin, thiamine, co- enzyme A** play a significant role in fatty acid metabolism of Cruciferae and oilseed crops like groundnut rotated with rice.
33. Known as **quick lime, slaked lime- CaO, Hydrated lime**- $CaOH_2$
34. **As a constituent of ferredoxin,** S participates in photosynthesis.
35. **Endemic species** – Native species for a specific area (Indigenous) – *Pheretima elongate, Lumbricus terestries, Allolobophora caliginosa etc.*
36. **Peregrine species** – Species established in countries other than (exotic) their normal homes –*Eisenia foetida* – Imported from Germany, *Eudrilus eugeniae.*
37. 1 t of Vermicompost is produced by – 2000 to 2500 earthworms in **4-8 weeks.**
38. **1800- 2000 worms/ m²** can convert 200 kg of waste in to compost in a month. 10 kg of earthworms (10, 000 worms approx.) occupying 5 m² area can convert one tonne of waste per month in to compost.
39. **One million earthworms** housed in a space of 22.6 × 22.6, 500 m² would convert 100 tonnes organic waste, upto 240 tonnes vermicompost in approx. one month.
40. **Crop removal**: 37.46 Mt of NPK (11.87 N + 5.27 P + 20.30 K Mt).
41. **Grain prod./kg NPK decreased** form 17kg in 1950-59 to 12 kg in 1960 – 69 and 9 kg in 1990-99.
42. **187.7 Mha** (57.1%) of total land area in India has been degraded and per capita arable land decreased from 0.34 ha in 1950- 51 to 0.16 ha in 1999.
43. **Food grain requirement**: 380 Mt by 2025 for 1.4 billion and 480 Mt by 2050 for 1.8 billion people.
44. **270-300 Mt** of organic manures of different kinds contributing 4 to 6 Mt of NPK. Negative balance of 8- 10 Mt of NPK leading to mining of soil reserves.

45. **About 25-30%** nutrient needs of Indian agriculture can be met by utilizing various organic sources.
46. **NO_3 inhibitor** = oxamide contain 31% N, Thiourea = 36.8% N, Dicyanadiamide = 42% N.
47. **P solubilizers** *(Bacillus, Pseudomonas, fungus)*.
48. **Phosphate absorbers** (Mycorrhiza).
49. **175 m metric tonnes** of N is fixed per year globally through biological N fixation.
50. **Legumes** contribute 40% of total biological N fixation.
51. **Pulses and oilseed** legumes occupy about 30 M ha, Requires **15000 t** of *Rhizobium* to cover the area, Current annual total production capacity of Rhizobium is 2685 t.
52. **Azolla** is a free floating water fern harbouring diazotrophic cyanobacteria *Anabaen aazollae* in its leaves responsible for N fixation.
53. **Predominant genera of BGA:** *Anabaena, Nostoc, Aulosira, Tolypothrix, Anabaenopsis, Calothrix, Scytonema, Fischerella, Stigonema, Campylonema.*
54. **Judicious use of BGA** could provide to the country's entire rice acerage as much N as obtained from 15-17 lakh tonnes of urea.
55. Availability of **Zn, Cu, K, S, Mg and Fe** also increases due to mychorrhizal activity.
56. Algae excrete **Vit.B_{12}, auxins and ascorbic acid.**
57. **Azolla** has very fast growing character and doubles its biomass in 3-5 days and fixes 40-60 kg N/ha.
58. **Every 100 g of *Azolla*** contribute 0.5g N, 0.5g P_2O_5, 0.4g Ca, 0.5g Mg, and 0.45g Fe.
59. **Nutrient content of Azolla** *pinnata* is 2.33 – 3.83, 0.196 and 0.153% N, P and K, respectively.
60. **50% N supplemented through Azolla** and incorporation of 10t/ha green biomass provide 25 kg N/ha after its decomposition.
61. State governments have issued **408 lakh Soil Health Cards** to the farmers up to **October, 2011.**
62. **Use of industrial wastes like pressmud**, urban wastes such as sewage sludge / rural wastes like fishpond effluent, crop residues and coir pith for composting augments nutrient sources for crops.

63. **Use of gypsum** (10-12 t ha^{-1}) for high pH sodic soils and lime (1-2 t ha^{-1-}) for acid sulphate soils improve use efficiency of several nutrients like **P, Zn, Ca and K.**
64. The area under very severe soil erosion (more than 40 tonnes per ha per year) constituted about **11 percent.**
65. **1 t of grain yield of rice** needs about 20 kg of N, 4 kg P and 25 kg K.
66. **Rain water may** provide about 5-10 kg N ha^{-1} per year all in nitrate form and to some extent S also (10-12 kg SO4 Ha^{-1}).
67. Soil is good for agriculture – **Loamy soil.**
68. Mineral has the lowest specific heat – **Quartz.**
69. Red colour of the soil – **Hematite.**
70. Amide fertilizers or Organic fertilizers – **Urea and calcium.**
71. N fertilizer is suitable in waterlogged condition – **Ammonium sulphate form.**
72. Weight of soil furrow slice – **2 x 10^6 kg/ ha.**
73. N percent in kisankhad – **20 – 25%.**
74. **Rice husk and burnt husk** (ash) are good sources of Sk.
75. **Phytin** has a high affinity for Zn and Fe and is considered more important in human nutrition.
76. Elements that provide structure to the plant – **C, H and O.**
77. Elements useful in energy storage – **N, S and P.**
78. Electron transport elements – **Fe, Mn, Mo, Cu, B and Cl.**
79. **Fe, Mn, Mo, Cu, B and Cl, known as** – Catalysers and activator.
80. Widely used as a soil conditioner – **polyacrylamide.**
81. **About 8 million tonnes** of plant nutrients are also washed away along with eroded sediments.
82. **Bacteria involved in P nutrition** – *Bacillus subtilis, B. megaterium, B. fluorescence, Pseudomonas* sp., *Escherichia* sp., *Xanthomonas* sp., *Achromobacter* sp., *Nitrosomonas, Thiobacillus thioxidans* etc.
83. **Fungi involved in P nutrition** – *Aspergillus awamori, A. niger, A. flavus, penicillium* sp., *Fusarium* sp., *Sclerotium rolfsii, Pythium* sp., *Rhizoctonia* sp. *Glomus* sp., *Gigaspora* sp., *Acaulospora* sp., (Endomycorrhiza) *Boletas, Amanita, Scleroderma* (Ecotomycorrhiza).
84. **Cellulolytic Fungi**: Chaetomium and Trichoderma.

85. **Cellulolytic Bacteria**: Cellulomonas and Clostridium.
86. **Cellulolytic Actinomycetes**: Nocardia & Streptomyces.
87. **Inoculation with *Azotobacter* and *Aspergillus*** along with rock phosphate is beneficial in improving the quality of compost in terms of N, C/N ratio, humus and available P.
88. **Primary colonizers** convert the complex carbohydrates and proteins into simple carbon sources (*Clostridium*).
89. **Secondary colonizers** break down organic acids into CH_4 and CO_2 (*Methanococcus, Methanobacillus and Methanobacterium*).
90. **Arsenic concentration** in ground water above 0.01 mg per litre is considered to be unsafe from the health point of view.
91. **Particulate pollutants**– Fumes, Aerosols, liquid droplets, pollen grains.
92. **Total viable count in** the range of 1 x 10^7 to 5 x 10^9 cells/g or ml of particular effective strain of the microorganism.
93. **Selenium contaminated** soil contain a selenium content many times higher than the permissible limit of 5 mg selenium per kg of produce.
94. **Gap of 10 million** tones of nutrients annually added and drained from the soil by crop removal and erosion.
95. **Green algae**: *Chlamydomonas, Chlorella, Codium, Dunaliella.*
96. **Diatom algae:** *Nitzchia.*
97. **Red algae**: *Ceramium, Corallina, Porphyridium.*
98. **Purple sulfur bacteria:** *Chromatium, Ectothiorhodospira, Thiocapsa.*
99. **Purple non sulphur bacteria:** *Rhodobacter, Rhodospirillum, Rhodopseudomonas, Rhodomicobium.*
100. **Green sulfur bacteria:** *Chlorobium, Pelodictyon.*
101. **Green non sulfur bacteria**: *Chlorofluxis.*
102. **Green manures** – Sunhemp (*Crotalaria juncea*), Dhaincha (*Sesbania bispinosa*), Cowpea (*Vigna unguiculata*), Cluster bean (*Cyamopsis tetragonoloba*).
103. **India generates.**
104. Rural compost : 285 million tonne/year.

    Crop residue : 273 million tonne/ year.

    City refuse : 14 million tonne/year.

105. The **commonly used earthworm species** are; *Eisenia foetida, Edudrilus eugeniae, Perionyx excavates, Dendrobaena veneta.*

106. **Japan imports 3000** million tonnes of earthworms per annum for waste conversion.

107. **Energy requirement for 1 kg fertilizer** is – 80 MJ/11.2 KWH for nitrogen, 12 MJ/1.1 KWH for phosphorus, 8 MJ/1 KWH for potash.

108. **Earth receives Rs. 580 thousand million** worth of N every year through biological sources.

109. **Rhizobial inoculation of chickpea**, lentil and soybean can generate an additional gain of Rs 11, 800 million from entire cropped area for which the inoculant cost will be Rs 35-40 million.

110. **BNF** – Lightening – 15 mt, Industrial – 50 mt, Microorganisms –175 mt.

111. **Legume crops** -35 mt, Non – legumes -09 mt, Permanent grasslands – 45 mt, Forest & woodland – 40 mt, Unused land – 10 mt, Sea – 36 mt.

112. **Average consumption of fertilizers in the country** currently at 144 kg per hectare of arable land (2010-11).

113. During 201-11, India imported 66 lakh tonnes of urea, 74 lakh tonnes of DAP and **64 lakh tonnes of MOP**.

114. In India, the average, soil loss rate has been calculated as **16.75 to/ha/year**, which translates into approximately 1 mm each year.

115. **Nitrogen Fixer** - *Rhizobium, Azotobacter, Azospirillum, Acetobacter*, BGA, *Azolla.*

116. **P Solubiliser** - *Bacillus, Pseudomonas, Aspergillus, Penicillium.*

117. **Plant Growth Promoting Rhizobacteria** – *Pseudomonas, fluorescence, Pseudomonas putida, Alcaligens, Azospirillum, Azotobacter.*

118. **NBS is applicable on DAP, MOP, MAP, TSP** and 13 grades of complex fertilizers.

119. Subsidy under NBS is same both for domestic and imported fertilizer products. NBS for SSP introduced w.e.f. **1st May, 2010**.

120. Import of all fertilizer products is under Opem General License (OGL) except **urea**.

121. N_2 fixation in pine trees: ***Azorhizobium***.

122. Lichens are the symbionts of the following organisms: **Fungus – Alga.**
123. **Lateritic soils** – Pedon is well developed horizon and eluviations + illuviation. Plinthite (Fe rich humus poor mixture of clays) formed.
124. Bacterial DNA is: **Double stranded linear with histones.**
125. Poly – hydroxybutyric acid is the reserved energy material that is equivalent to lipid is present in: **Bacteria.**
126. Cold sterilization agent: **Liquid nitrogen.**
127. The common igneous rock found in India is : **Granite.**
128. India consumption **~15% of global Nitrogen,** India consumption ~ **19% of** global Phosphate, India consumption ~ **15% of global Potash.**
129. Nutrient Demand growth rate: **N ~ 3%, P ~ 6%, K ~ 5%.**
130. Applicable rate of Import (Customs) duty is 5% on fertilizer products.
131. **Saline < 15> 4 < 8.5, Saline alkali > 15> 4> 8.5, Alkali/sodic> 15 < 4 > 8.5.**
132. **Soil less media** – Peat, Sphagnum Moss, Vermiculite, Perlite, Leaf Mould Cocopeat, Shredded Bark, Sawdust and Wood Shavings.
133. The rate of flow is in the following sequence. **Sand > loam > clay.**
134. Advanced methods of cultivation – Hydroponics, solution culture, rock wool **culture, Nutrient film technique.**
135. The rate of unsaturated flow is in the following order. **Sand < loam < clay.**
136. Which plant nutrient is leached maximum? **Ca.**
137. Well decomposed manure should have C: N ratio – **10:1.**
138. Best time for orchard soil culture is – **Nov – Jan.**
139. Soil erosion in orchard can be checked by- **Sod culture.**
140. Maximum heating is obtained with which mulch material – **Black.**
141. An example of slow release fertilizer is – **Coated urea.**
142. D leaf is a best indicator of nutrient status of – **Pineapple.**
143. Features of cold desert – **Heavy influx of IR & UV light, Low air density.**
144. Soil erosion loss in NEH region is to the tune of --- t/ha/ year **-40.**
145. Nutrient highly mobile in plant is – **N.**

146. Optimum soil pH for phosphorus availability to plants is – **6.5- 7.5.**
147. NDSSLUP has delineated the country into how many agro ecological regions – **21.**
148. North Eastern Hill Region has which soil type – **Red laterite.**
149. Highly shade tolerant fruit crop is – **Carambola.**
150. Highly drought tolerant fruit crop is – **Ber.**
151. Fruit crop coming well in acidic soil is – **Strawberry, Raspberry, Gooseberry.**
152. High C: N ratio favours – **Flower development.**
153. Nitrogen deficiency symptoms are first visible on – **Older leafy.**
154. Phosphorus deficiency retards – **Growth, Induces late maturity, Purple colouration of leaf.**
155. P in excess causes deficiency of – **Fe, Zn, K.**
156. More calcium is present in – **Older leaf.**
157. Excess of manganese may at times give symptoms of – **Fe deficiency.**
158. Excess sodium causes – **Ca deficiency.**
159. In acidic soils which element is at toxic level – **Al.**

Essentiality of the elements established by

N	Theodore De Saussure (1804)
C	Priestly
Ca, Mg, K, S	Carl sprengel (1839)
P	Von Liebig (1844)
Fe	E. Greiss (1844)
Mn	J.S. Hargue (1922)
Zn	Sommer and Lipman (1926)
Cu	Sommer, Lipman and Mc Kenny (1931)
Mo	Arnon and Stout (1939)
Na	Brownell and wood (1957)
Co	Ahamed and Evans (1959)
B	Warring ton (1923)
Cl	Broyer (1954)

Agro – Chemicals

160. Insecticide for termite - **Chlorpyiphos.**
161. Eco-friendly insecticides - **Neem products, microbial pesticides (BT, GV, NPV).**
162. Pheromone trap attracts - **Male moths.**
163. Chlorinated hydrocarbon insecticides - **Endosulfan, BHC, Dicofol, DDT.**
164. Systemic insecticides of organophosphate - **Monocrotophos, Phosphamidon, Dimethoate, methyleDemeton/ OxydemetonMethyle and Ethion.**
165. Selective insecticides - **Endosulfan, phosalone, Malathion and Menazon.**
166. Carbamate insecticides - **Carbofuran, Propoxur and carbaryl.**
167. Insecticides synthetic pyrithroids - **Fenvalerate, Cypermethrin, Decametrin & Permethrin.**
168. Insecticide which is being opposed by environmentalist is - **Dichlorvos (DDVP).**
169. Insecticide used in mosquito mats is - **Allethrin.**
170. Insecticide has anticholinesterase action on insect nervous system - **Malathion.**
171. In storage, grains can be mixed with - **Malathion dust (5%).**
172. An alleochemical beneficial to releaser only is called - **Allomone.**
173. Kairomone is beneficial to - **Receiver.**
174. Synomone is beneficial to - **Receiver & Emitter.**
175. Antimones are non advantageous to - **Both Receiver & Releaser.**
176. Carbamates found in - **dimoseb.**
177. Country which has highest consumption of herbicide - **Japan.**
178. Used in plastic mulch - polyethylene, polyvinyl chloride.
179. Seed hardening chemical is used in sorghum and bajra - **KH_2PO_4.**

180. Seed hardening chemical is used in maize - **Kcl.**

181. Seed hardening chemical is used in cotton - **Cycocel / succinic acid.**

182. Seed hardening chemical is used in groundnut - **$CaCl_2$.**

183. Seed hardening chemical is used in pulses - **Ash.**

184. Used as fumigant - **methyl bromide.**

185. Improve the colour of apple - **MCPA.**

CHAPTER 5

Physiology

1. Low of tolerance was introduced by - **Shelford.**
2. End product of glycolysis is – **Pyruvate.**
3. Define **lithophytes** - The plant grown is rock cervices are called lithophytes.
4. During glycolysis , the Net gain of ATP is - **2.**
5. For reclamation of saline and Alkaline soils used are - **Gypsum and Pyrite.**
6. The term ecosystem was coined by - **A.G. Tansley (1935).**
7. What is **parthenocarpy** - The development of fruit without fertilization.
8. The development of embryo without fertilization is called - **Apomixes.**
9. Relative humidity is measured by- **Pshychrometer.**
10. Father of Indian pale botany is - **Birbal Sahni.**
11. The weed used for ornamental purpose is - **Lantana Camaera.**
12. The core component of chlorophyll is - **Mg.**
13. The soybean is sown at the depth of - **3 cm.**
14. Chaisma is seen during - **Diplotene sub - Phase.**
15. Photo- Respiration occurs in - **Chloroplast.**
16. Water use efficiency is highest in which plants - **CAM Plants.**
17. Chromosomal theory of inheritance was proposed by- **Sutton & Boveri.**
18. Which chemical is mostly used for breaking tuber dormancy in potato-**thiourea.**
19. The term water harvesting was first time used by - **Mayers.**

20. The first commercial hybrid cotton of the world is - **Hybrid -4.**
21. The number of electrons for convertion of No_3 to NH_4-**16.**
22. Dormancy breaking hormone is - **cytokinin.**
23. The element involved in energy transfer and storage in plants is - **Phosphorus.**
24. Fruit ripening hormone is - **Ethylene.**
25. Luxury consumption or maximum uptake nutrient is- **K.**
26. Crop canopy temperature is measured by - **Infra red thermometer**
27. Secondary Plant nutrients are- **Ca, Mg, & S.**
28. Inflorescence of sugarcane is called - **Arrow.**
29. Shedding of plants part is due to - **ABA hormone.**
30. Chemical used for delinting of cottons - **H_2SO_4.**
31. Movement from higher concentration to lower concentrationn is known as - **diffusion.**
32. Who discovered vitamin? **Funk.**
33. Lodging preventor is - **CCC (Cycocel).**
34. Root promoting hormone is **IBA.**
35. Who discovered mitocondria? **C. Benda (1897).**
36. The term cytokinin was proposed by - **Letham (1963).**
37. Kinetin and zeatin are the - **Cytokonins.**
38. photosynthesis is an - Oxidation - **Reduction process.**
39. Gibberellin was discovered for the first time used by - **Kurosowa.**
40. Water soluble vitamin is - **Vitamin C.**
41. Acidity in gram leaves is due to - **Malic acid / oxalic acid.**
42. Photosynthetic rate is highest in **C_4 plant.**
43. Lock and Key Model was proposed by - **Fisher.**
44. Anti - Sterility vitamin is - **Vitamin E.**
45. Netural parthenocarpy is found in - **Banana.**
46. The term vernalization was coined by - **T.D. Lysenco.**
47. Chemical relationship is done between two cells by-**Plasmodesmata.**
48. Tetrazolium Test is done for - **viability and vigour.**
49. The cytoplasm and nucleus combinedly called - **protoplasm.**
50. Cell theory was proposed by **M.J. Schleden and Theodo Schwann.**
51. The term Lysosomes was first used by **Dave in 1955.**

52. The main function of Golgi bodies are - packaging of food materials such as **protein, lipids etc.**
53. The function of ribosomes are - **protein synthesis.**
54. The term mitochondria was given by - **C. Benda (1897).**
55. The simple microscope was invented by - **Galileo (1610).**
56. The shortest phase of all mitosis phase is - **Anaphase.**
57. Meiotic division is also known as **reductional division.**
58. The term mitosis was coined by **Fleming (1882).**
59. Mitosis occurs in **Somatic cells.**
60. Segregation occurs during - **meiosis.**
61. The longest mitotic phase is -**Prophase.**
62. Absorption Spectrum of photosynthesis is **blue and red.**
63. Action Spectrum of photosynthesis is **red and blue.**
64. Dark reaction was discovered by **Blackman.**
65. Dark reaction is also known as - **Blackman reaction or C3 – cycle.**
66. The main site for dark reaction of photosynthesis is – **stroma.**
67. The evaporation of water from plants is called - **transpiration.**
68. Evaporation from soil and transpiration combined are called **ET.**
69. C4 Plants proposed by **Hatch and slack (1970).**
70. The dark reaction process of photosynthesis has been named variously - **Calvin cycle or Blackman reaction.**
71. Light reaction is also known as **Hill reaction.**
72. In photosynthesis, the solar energy converts into - **Chemical energy.**
73. The OSMOSIS term was given by - **Abble Nollet (1748).**
74. The term "diffusion pressure deficit " was introduced by - **Meyer in 1938.**
75. The term Imbibition was coined by - **Sachs.**
76. Imbibition Pressure is also known as - **Matric Potential.**
77. Root pressure word was coined by- **Stephan Hales.**
78. Father of Physiology - **Stephan Hales.**
79. Soil les cultivation of plants is known as **Hydroponics.**
80. Rate of transpiration is determined by - **Potometer.**
81. The loss of sap (water) from the injured parts of plants is called - **Bleeding.**

82. Approximately 90% of the world photosynthesis is carried out by - **Marine and fresh water algae.**
83. Growth rate is measured by **Auxanometer & Cresco graph.**
84. Photoperiodism term was coined by **TD Lysenko (1920).**
85. Apical Bud dominance is caused by which hormone **Auxin.**
86. Which hormones is used as a herbicide? **auxin.**
87. Which hormone is related to drought tolerance? **absicic acid (ABA).**
88. Dormancy break by which hormone **Cytokinis.**
89. Stress hardening in plants can be active by which hormone **ABA.**
90. Fruit ripening hormone is **ethylene.**
91. The term auxin was used by **Frits Went Holland (1905).**
92. Want's auxin is now known as **Indole -3-acetic acid (IAA).**
93. Which hormones are the Auxin **IAA, IBA, NAA, 2-4-D,2,4,5-T' MCPA, IPA TIBA.**
94. Shedding of plant parts is due to **ABA hormones.**
95. Cycocel (CCC) is a **growth retardant.**
96. The most widely used auxin herbiceds are **2-4-D, 2,4,5-T' MCPA.**
97. Gibberellins were first discovered in **Japan.**
98. The hormones also known as **Photohormones, growth hormones, growth substances.**
99. Main auxin is also known as **IAA.**
100. Root promoting hormones is **IBA.**
101. The precursor of IAA is **Tryptophane.**
102. Synthesis of IAA in plants requires **Zinc.**
103. The term phytohormone for hormones of plants is suggested by **Thimann (1948).**

OOO

CHAPTER 6

Agricultural Entomology

1. PDBC is located at **Bangalore.**
2. The major protein found in silk is **Fibroin.**
3. NCIPM is located at **New Delhi.**
4. Corbiculate type of legs are present in **Worker honey bees.**
5. Forewings are present in **Mesothorax.**
6. A median muscular pad present in between a pair of lateral claws in the foot of insects is called as **Arolium.**
7. Hyper pneumatic type of tracheal system is found in **Diplura.**
8. Naiads of Anisoptera have **Rectal gills.**
9. Johnston's organs are present in **Pedicel** part of the antenna.
10. In cicadidae, tympanal organs are present in **1st abdominal segment.**
11. Gustatory receptors are associated with **taste.**
12. In male cicada, tymbal is present in **1st abdominal segment.**
13. In fireflies, light is produced due to oxidation of **Luciferin.**
14. The only cellular layer of integument is **Epidermis.**
15. The layer which separates epidermis from the haemocoel is **Basement membrane.**
16. Moulting fluid during moulting is secreted by **Epidermis.**
17. Non-chitin as part of the cuticle is **Epicuticle.**
18. The layer which serves as a water proof layer of the integument is **Wax layer.**
19. Nitrogenous polysacharide present in insect integument is **Chitin.**
20. Insect head consist of **6 segments.**
21. Technical term of head capsule is **Cranium.**

22. The line of weakness is present in /on Head region called as **Epicranial suture.**
23. Below clypeus we find **Labrum.**
24. Lateral side of head is called **Gena.**
25. Compound eyes are consist of smaller eyes called as **Ommatidia.**
26. Ocelli in cockroach are called as **Fenestrae.**
27. Primitive or Basic type of mouth part is **Biting and chewing.**
28. A pair of upper jaws present in mouth parts these are called as **Mandibles.**
29. Thorax is made up of **3 segments.**
30. Segmental plate which is present on dorsal side of thorax is called as **Tergum / Notum.**
31. Lateral plate of thorax is called as **Pleuron.**
32. Leg consist of primitively **6 segments.**
33. First tarsal segment is called as **Basi tarsus.**
34. Fore wings of cockroach are called as **Tegmina.**
35. Insects which do not undergo metamorphosis are come under the group **Ametabola.**
36. Hemimetabolous development is also called as **Indirect metamorphosis.**
37. Some immature insect are able to reproduce is called **Paedogenesis**
38. Outer layer of egg is called **Chorion.**
39. The shape of eggs of Bugs is **Barrel.**
40. Puparium is nothing but **Co-arctate.**
41. Butterfly pupae are called as **Chrysalis.**
42. Typical antennae contain **3 parts.**
43. Johnston's organ present in **Pedicel.**
44. Cockroaches have type of antennae called **Setaceous type.**
45. Male Mosquitoes contain **Plumose type of** antennae.
46. Aristate type of Antennae is present in **House fly.**
47. The thickest part of the leg **Femur.**
48. Last part / segment of leg is called **Pre tarsus.**
49. Fossorial type of legs means **Digging type.**
50. In grass hopper which legs are modified for jumping purpose **Hind legs.**

51. In Antennae cleaner type of legs the modified parts are **Tibia and Basitarsus.**
52. In pollen carrying type the modified parts of legs are **Tibia and Basitarsus.**
53. Plume moth have **fissured wings.**
54. Fringed wings present in **Thrips.**
55. Elytron type of wings present in **Beetles.**
56. Eyes which are involve in only regulation of light intensity **Simple eyes.**
57. Compound eyes which are separate each other, condition is called as **Dioptic eyes.**
58. For construction of comb, honey bees uses **Mandibles.**
59. Roya jelly is secreted by **Hypopharyngeal glands.**
60. In honey bees the meta thorax is fused with the I abdominal segment called **Propodium.**
61. Rasping and sucking type of mouth parts present in **Thrips.**
62. Insects which jump without using any legs **Maggots of flies.**
63. Carbiculum is modified for **Pollen basket.**
64. Ptero thorax made up of **Meso and Meta thorax.**
65. Wing flexion mechanism well developed in **Neoptera.**
66. Jugate and frenate type of wing coupling mechanism is seen in **Moths.**
67. Amplexiform type of wing coupling mechanism is seen in **Butterflies.**
68. Frequency of wing beat is expressed in terms of **Hz/ sec.**
69. Outermost layer of body wall is **Cement layer.**
70. Among these layers of body wall which is the cellular layer **Epidermis.**
71. Cement layer is the part of **Epicuticle.**
72. The protein which is involved in hardening of cuticle **Sclerotin.**
73. For wing articulation and for musule attachment the protein involved is **Resilin.**
74. Moulting fluid is produce from **Epidermis.**
75. Body processes/appendages which are incapable of movement **Spines.**

76. Pheromones are used for **Intra specific communication.**
77. In thrips, asymmetrical type of mouth part is because of **Right mandibles are absent.**
78. Some insects do not have malpighian tubules like **Collembolans and Aphids.**
79. In embioptera silk glands are present on **Basitarsus.**
80. Similar form of wings are fond in **Isoptera.**
81. In strepsiptera Fore wings are **reduced.**
82. Corium is present in **Egg.**
83. **Excretory system** is responsible for Homeostatic.
84. Malpighian tubules absent in **Aphids and Collembola.**
85. Wide spread excitatory transmitter is **Acetylcholine.**
86. Adults live for few hours in **Ephemeroptera.**
87. Male genitalia in odonata present on **2 and 3 abdominal sterna.**
88. Fenestrae present in **Cockroach.**
89. Semicircular hind wings are seen in **Earwigs.**
90. In Nepidae hind legs modified for **Swimming.**
91. Carinae is present in **Lophopidae.**
92. In Hover flies spurious vein present between **R & M.**
93. Bifid claw present in **Pieridae.**
94. Number of recurrent veins present in Ichneumohidae **2.**
95. Hyper metamorphosis is seen in **Strepsiptera.**
96. Anopheles mosquito eggs are **singly flaoting.**
97. Non chitinous layer of integument in insects is **Epicuticle and Endocuticle.**
98. The layer of Epicuticle, which prevents water loss is **Lipid layer.**
99. Pore canals are present in which layer of the insect integument **Epicuticle.**
100. Thickest layer of the integument is **Exocuticle.**
101. Moulting fluid is produced by **Hypodermis layer** of the integument.
102. Wooly aphid is the serious pest of **Sugarcane.**
103. The first Bt transgenic crop under field testing in India is **Cotton.**
104. Which is absent in arthropoda? **Closed blood vascular system.**
105. Prickly pear is eradicated by the insect ***Dactylopius tomentosus.***
106. Hopper burn symptom is due to **Leaf hopper.**

107. The stem borer which is common to sorghum, maize, rice and wheat is ***Sesamia inferens***.
108. Curling of leaves in chillies is caused due to **Thrips**.
109. Royal jelly is rich in **Proteins**.
110. Major pest of chickpea is **American bollworm**.
111. Silk is produced by the caterpillars of **Bombycidae and saturniidae**.
112. 65% of the world's lac production is in **India**.
113. First fused abdominal segment of hymenoptera is known as **Propodeum**.
114. Carmine red dye is extracted from **Cochineal insects**.
115. Pollen basket is formed by ***Hind tibia***.
116. The Italian bee introduced into India is ***Apis mellifera***.
117. Sting of the honeybee is modified **Ovipositor**.
118. In mulberry silkworm, silk is secreted by the larvae from **Labial glands**.
119. The highest lac production is in the state of **Bihar**.
120. Malic and oxalic acids are collected from green leaves of **Bengal gram**.
121. The organ that helps in excretion and regulation of the internal environment in insects by controlling ion and water balance in blood is **Malphigian tubules**.
122. Lac is marketed as **Shellac**.
123. Functional female of the bee colony is **Queen**.
124. Largest size of the honey bee is ***Apis dorsata***.
125. The forewings of stylops are called **Halteres**.
126. Harvest of mangoes is more difficult due to presence of **Mealy bugs**.
127. A regular pest of Okra is **Fruit borer**.
128. A persistent pest occurs on a crop almost for **Twelve months**.
129. Mustard sawfly belongs to the order **Hymenoptera**.
130. "Katte" disease of cardamom and Bunchy top of Banana are transmitted by **Aphid**.
131. A major pest of tomato is **Fruit borer (*Helicoverpa armigera*)**.
132. Trichogramma is a **Egg parasitoid**.
133. On which stage of rice the Gundhi bug feeds **Milky stage**.
134. A white grub is a **Chafer beetle**.

135. The name of rice yellow stem borer ***Scirpophaga incertulus***.
136. The highest consumption of pesticide in India in the crop **Cotton.**
137. Cotton whitefly belongs to **Aleyrodidae.**
138. Double seed formation is due to **Pink bollworm.**
139. Destructive Insect Pest (DIP) act was passed in the year **1914.**
140. The insecticides act was passed in **1968.**
141. The Central Plant Protection Training Institute is located at **Hyderabad.**
142. The safest insecticide for honeybee is **Endosulfan.**
143. Oesophageal ganglion in insects is otherwise known as **Hypocerebral ganglion.**
144. What is the sequence of coating of seed by insecticide, fungicide and rhizobium? **Fungicide + insecticide + rhizobium.**
145. Dead heart in young seedlings of wheat is caused by ***Sesamia inferens***.
146. Function of the Mandibular glands is **To soften the wax.**
147. Silver shoot in rice is caused by ***Orseolia oryzae***.
148. Rice mite is ***Oligonychus oryzae***.
149. Bromadiolone is an effective **Rodenticide.**
150. Eggs are laid in an ootheca by **Cockroach.**
151. *Aedes aegypti* is a vector of **Dengue.**
152. Major pest of furniture is **Powder post beetle.**
153. Housefly spread the disease **Cholera.**
154. Sand flies transmit the disease **Kala azar**.
155. Stem girdler is a pest of **Grape.**
156. Major pest of pomegranate is **Fruit borer.**
157. Major pest of sapota **Seed borer.**
158. Fruit sucking moth can be controlled by **Bagging of fruits.**
159. Black fly is a serious pest of **Citrus.**
160. Major pest of cashew is **Tea mosquito bug.**
161. An introduced pest from Italy **Potato tuber moth.**
162. Diamond back moth is a pest of **Crucifers.**
163. Cucumbers major pest is ***Bactrocera cucurbitae***.
164. Serpentine leaf miner ***Liriomyza trifolii***.
165. Brinjal shoot and fruit borer is ***Lucinodes orbonalis***.

166. Blue butter flies are the pests of **Pulses.**
167. *Earias vittella* is also found on **Okra.**
168. Stem galls in cotton is due to **Stem weevil.**
169. ETL for cotton bollworms in Central India is **10% damage.**
170. One generation is completed in a year in **White grubs.**
171. *Goniozus nephantidis* is a parasitoid of **Black headed caterpillar.**
172. *Uroleucon carthami* is a serious pest of **Safflower.**
173. This is the only Hymenopteran pest belonging to family **Tenthredinidae.**
174. *Acherontia styx* is a pest of **Sesamum.**
175. It is now fruit sucking moth of citrus ***Achaea janata.***
176. Egg masses collection is the best control method for **Red hairy caterpillar.**
177. NPV is effective in the control of ***Helicoverpa armigera.***
178. *Dipha aphidivora* is an effective predator of **Wooly aphid.**
179. Dead hearts within 28 days of germination in sorghum are due to **Stem borer.**
180. Sex of the cockroach is recognized by the presence of **Cerci.**
181. In winged insects meso and metathoracic segments are collectively known as **Pterothorax.**
182. Pulses are attacked both in the field and storage by ***Calloso bruchus.***
183. Johnston's organ is located on **Pedicel.**
184. Cervical sclerites are present in **Neck.**
185. Filter chamber is present in **Digestive system.**
186. In winged insects, dilated trachea are known as **Air sac.**
187. Only the prothoracic spiracles are open **Propneustic.**
188. Outer wall of the egg is known as **Chorion.**
189. Nymph of dragonfly is known as **Naid.**
190. Pterostigma and archedictyon wing venation is the characteristic feature of **Odonata.**
191. Sub-image stage is the feature of order **Ephemeroptera.**
192. Assasin bug belongs to the family **Reduviidae.**
193. Asymmetrical mouthparts are found in **Thrips.**
194. Family Mantispidae belongs to the order **Neuroptera.**
195. Antilion flies are included in the order **Neuroptera.**

196. Glow worms belong to the order **Coleoptera.**
197. Haemoglobin is found in **chironomid larva.**
198. The smallest known insect in class insecta is **Fairy fly.**
199. Crochets are found on **Abdominal legs.**
200. Cryptonephridial condition is found in **Stored grain pests.**
201. Only Lepidopteran parasitoid on Pyrilla ***Epipyrops melanoleuca*.**
202. Minute pirate bugs belong to family **Anthocoridae.**
203. Which crystalline protein is used in MECH-12 cotton hybrid **Cry-1AC.**
204. CABI is known for **Identification service.**
205. Directorate of Plant Protection, Quarantine and Storage is situated at **Faridabad.**
206. The period in between two moults in insects is known as **Instar.**
207. Sensory organs are provided with **Trichogen cell.**
208. Enzymes concerned with digestion in insects are mainly present in **midgut and salivary glands.**
209. Insects feeding on stored products generally have higher activity of enzyme, **amylase.**
210. Some extra-intestinal digestion in insects occur in **predaceous beetles.**
211. As adults, female insects have a higher need than males for **dietary proteins.**
212. Ambrosia beetles are associated with fungi that enable them to **use xylems of woody plants.**
213. Heart is restricted to the insect's posterior abdominal segments only in **lice.**
214. In most insect species heart is innervated by nerves running from the segmental ganglia, but not in **Anopheles.**
215. The antennae are used for mating in **collembola.**
216. The mandiblies in adults are vestigial or absent in **Ephemeroptera.**
217. The labium is known as rostrum in **Hemiptera.**
218. Biting and chewing insects have contact chemorecptors on all the mouth parts except **mandibles.**
219. Blood sucking insects that do not require the presence of a nucleotide to take a full blood meal are **sandfly.**

220. Insects feeding on the internal fluids of plants, secrete a sheath around the stylets in **homoptera**.

221. Factor X and thrombin are chemicals in vertebrates that regulate coagulation. But peptides in insects that counteract effect of both chemicals are present in **saliva**.

222. The mutual or unilateral exchange of alimentary fluids including saliva is called **trophallaxis**.

223. Stretch receptors differ from all other mechanoreceptors in consisting of **multiplolar neuron**.

224. In crickets, tympanal organs of legs receive sound **via trachea and spiracles**.

225. Gyrinid beetles perceive ripples on the water surface through **echolocation**.

226. The only visual organs of larval holometabolous insects are **stemmata**.

227. The degree of fineness with which an eye forms an image of an object is referred to as **resolution**.

228. Neurosecretory cells normally occur in the central nervous system in the **ganglion**.

229. Social insects regulate the temperature of their nests so that **larvae develop optimally**.

230. Nephrocytes are cells in the haemocoel that take up chemicals of relatively **high molecular weight**.

231. Haemolymph of fresh water insects is **hypertonic**.

232. In insects production of primary urine and resorption are **independent**.

233. Gas that is highly toxic and is the primary nitrogenous end product of protein metabolism in insects is **NH_3**.

234. The three distinct regions in malpighian tubule are made up of **different cell types**.

235. Some insects have structures to hold a thin film of air on the outside of body for gaseous exchange and film of gas is called **plastron**.

236. Cutaneous respiration is present in insects like **protura**.

237. Delay in development evolved in response to regularly recurring periods of adverse environmental condition is **diapause**.

238. The larval development in most insects is **Epimorphic**.

239. In 2002, a new insect order named was **Mantophasmatodea.**
240. The concentration of a 14 SC indoxacarb in a spray solution if 0.5 ml is used per litre of spray soluition is **0.007%.**
241. The principal active moiety in neem is **Azadirachtin.**
242. Fipronil belongs to the group of **phenyl pyrrozole.**
243. Imidacloprid insecticide belongs to the group of **chloronicotinyl compound.**
244. Journal with secondary source of information on different aspects of entomology is **Review of Applied Entomology.**
245. In 1944, Gerhard schrader developed systemic insecticide schradon which is **organophosphate.**
246. At low concentrations, neem prolongs larval period because of **high JH activity.**
247. Bt. toxins are non-toxic to higher animals because of no dissolution of cry toxin in view of pH, which is **acidic.**
248. Bt toxins are **protoxin.**
249. In cyclodine compounds , endrin was banned because it is highly toxic to **snakes in paddy fields.**
250. DDVP has a slight fumigation, contact and **stomach poison.**
251. Monocrotophos, generally, is not advocated for pest suppression on vegetables **because of Phytotoxicity and residual effect.**
252. Predaeceous beetle with head wider than pronotum **cicindellidae.**
253. Polymorphism is characteristic of **termites.**
254. Insect killer of salvinia weed is **Cyrotobagous.**
255. Antennae cleaner in honey bee is in the **middle leg.**
256. A hyperparasite is a **parasite on parasite.**
257. Covering of tomato nursery with 200 mesh mosquito net is recommended to check ***Bemesia* sp.**
258. *Asphondylia sesami* causes **galls in flower buds.**
259. NPV is expressed as **PoBs/ml.**
260. To manage sugarcane borer, Trichogramma parasitoids can be released at **1.6 lakh/ac.**
261. Sir Ronald Ross's famous studies associated with the discovery of transmission of malarial parasites by Anopheles were conducted at **Sikandarabad (U.P.).**

262. The larvae of insects that exhibit adenotrophic viviparity spend most of their time **inside womb of mother.**
263. Secondarily wingless adult dipteran can be identified by **enlarged mesothorax.**
264. Mullerian mimicry is **an unpalatable species mimicking another unpalatable species.**
265. Recently deaths of birds were reported due to baits for suppressing caterpillar pests on pulses because **bait pellets were applied during morning hours.**
266. Microbial pesticides are effective against target pest insects when field conditions are **humid.**
267. A new insect pest on blackgram and greengram reported from parts of north Karnataka is **Apion.**
268. Light traps are useful in surveillance, monitoring and suppression of insect pests and the source of light should preferably be **UV bulb.**
269. For effective suppression of thrips on sunflower, Imidacloprid can be recommended at **0.3 ml/litre.**
270. A pest on groundnut that underwent annually two generations, but currently has overlapping generations ***Amsacta* sp.**
271. Invasive gall inducer in nurseries of Eucalyptus is ***Leptocybe* sp.**
272. Redheaded hairy caterpillars prefer certain wild plants over cultivated crops. The plants are **Calotropis and Jatropha.**
273. Castor can be planted as a border crop in groundnut to serve as a trap crop against ***Spodoptera* sp.**
274. For effective management of cereal, lepidopterous stem borers, application of insecticides is recommended in **whorls.**
275. Insect cannot synthesize **steroids.**
276. Merthyl Bromide is being gradually phased out against stored product pests because **affects ozone layer in stratosphere.**
277. Flowers have beauty only honey bees can see because of **perception of both UV + IR light rays.**
278. Nymphs live in spittle on plants **cercopidae.**
279. Pollens carried on underside of abdomen **megachilidae.**
280. Growing mustard as a trap crop is recommended in cabbage to check the infestation of **Plutella.**

281. A kairomone is a chemical benefecial to **Receiver.**
282. Example for animal derived insecticide is **Neries.**
283. Bollgard marketed by Monsanto is **a gene with bacterial protein against caterpillars.**
284. Host plant resistance was propounded by **R.H. Painter.**
285. The function of urogomphi on certain larval coleopteran abdominal segments serve the function of **defense.**
286. The whole process of spermatogenesis in adult insect is completed before **ecolosion.**
287. Aphids release an alarm pheromone from the cornicles when **attacked by parasites or predators.**
288. Dr. Paul A. Hedin, USDA has written a book on **Host Plant Resistance.**
289. Molecular markers hold good promise in resistance breeding such as **restriction fragment length polymorphism.**
290. Chemical stimulants of Brassica for ovipositional preference by Plutella are **glucosinolates.**
291. A bioassay technique to establish the potential of plant volatiles in olfactory stimulation of an insect is **EAG.**
292. The two statistical parameters to study spatial distribution of insects in space are **Mean and variance.**
293. The maximum concentration of pesticide residues that is legally permitted in food commodities is called **Maximum residue limit.**
294. A biocontrol agent of Lantana weed in India is **Teleonemia.**
295. Polyphagous spider mite affecting a wide range of crops belongs to genus **Tetranychus.**
296. Wormiform eriophyid mite in southern India which devastated plantations of **Coconut.**
297. Synthetic analogue of DDT insecticide exclusively toxic to most of the phytophagous mites **dicofol.**
298. The predominant excretory end product of protein breaks down in terrestrial insect is **uric acid.**
299. Until 2002, the most recent insect order was Grylloblattodea described in 1914. But in 2002, a new order was described with the name **Mantophasmatodea.**
300. An important object of the ICZN is to provide **uniqueness, stability and distinctiveness.**

301. The two types of compound eyes found in insects are **Apposition and Superposition type**.
302. Storage excretion is the function of **Urate cells.**
303. Sub-social insects living in silken tunnels are **Embiids.**
304. The larva construct pit fall traps in **Ant lions.**
305. Phase polymorphism is shown **Grasshoppers.**
306. Plastron respiration is observed **Aquatic insects.**
307. Typhus is transmitted by **Body louse.**
308. A nozzle suitable for herbicide application is **Flat pan.**
309. We can get the best spray deposit on plant by using **Mist blower.**
310. An extremely toxic insecticide will have LD 50 of **1-50 mg/ kg.**
311. An insecticide safer to honey bees is **Phosalone.**
312. Acyl urea compounds are **Chitin inhibitors.**
313. An insecticide that causes resurgence of spider mites is **Deltamethrin.**
314. Avermectins are derived ***Streptomycis* sp.**
315. *Cryptolaemus* sp. is mass multiplied in the laboratory on **Mealy bugs.**
316. The weed killer for the suppression of *Chromolaena odorata* is ***Procedochares* sp.**
317. *Chrysoperla* sp. is mass multiplied on the **egg** stage of rice moth.
318. A toxin isolated from marine annelid is **Nereistoxin.**
319. Methyl eugenol is a food attractant for **Oriental fruit fly.**
320. The major detoxifying enzymes involved in resistance to pyrethroids are **MFO.**
321. Two important entomological journals issued by the Entomological Society of America Are: **Journal of Economic Entomology and Environmental Entomology.**
322. The predominant sugar in insect haemolymph is **Trehalose.**
323. The vitellogenesis synthesis in fat body is stimulated by **JH.**
324. *Opisina arenosella* usually lays eggs **near old larval galleries.**
325. Drooping of coffee berries with a hole in the naval region is due to ***Hypothenemus hampei*.**
326. In cardamom plantation, burning of larvae by using fire torches is recommended to manage ***Eupterota* sp.**

327. What is not a measure of diversity **Gausse's index**.
328. Male and female produced complex epideictic pheromones are a feature associated with **Scolytidae.**
329. Bunchy top in sugarcane is caused by ***Scirpophaga* sp.**
330. A *Bt.* subspecies that are specific to maggots and caterpillars are ***isralensis* and *kurstaki*.**
331. The following are the predatory Lepidoptera found in Karnataka ***Dypha, Epipyrops, Spalgis*.**
332. The medium used to clear specimens such as mites before mounting on slides is **Hoyer's medium.**
333. Vasiform orifice, ctenidia, labellum and strigil are the features associated with the following taxa in that order. **Whiteflies, fleas, houseflies, Apocrita.**
334. How is a scientific name written correctly? ***Macropsis lalithae* Viraktamath, 1979.**
335. Navigation in honey bees is facilitated through: **Sun compass, Polarized light and Landmarks.**
336. One of the phenomena observed in insects that allows for eternal effectiveness of insecticides in pest management is **insecticidal negative cross resistance.**
337. Larval body shape in many holometabolous insects such as caterpillars is maintained largely due to **Hydrostatic pressure.**
338. Two principal nitrogenous excretory products of insects are **Ammonia and Uric acid.**
339. Monophagous insect pest **Yellow stem borer of rice.**
340. A Lady Bird Beetle is a **predator.**
341. Honey dew is excreted by **Aphids (homopteran insects).**
342. Petriophic membrane is found in **Mid gut.**
343. Ecdysone is secreted by **Prothoracic gland.**
344. Brown plant hopper is a major pest of **Paddy.**
345. Nuclear polyhedrosis virus is most effective against **Lepidopterous insects (caterpillars).**
346. National Centre for Integrated Pest Management is situated in **New Delhi.**
347. In which year the Directorate of Plant Protection, Quarantine and Storage was established? **1946.**

348. Honey bees most suitable for Apiculture? ***Apis mellifera***.
349. Dark yellow colour on the label of container of pesticide indicates that pesticide is **Highly toxic**.
350. Cyst nematode belongs to genus ***Heterodera***.
351. Flacherie disease in silkworm is caused by **Bacterium**.
352. Scale insects belong to the order **Hemiptera**.
353. Use of synthetic pyrethroids has caused serious outbreaks of **Whitefly**.
354. Cryptonephridial condition occurs in **larvae and adults of Coleoptera and larvae of Lepidoptera**.
355. The common name of *Chilo partellus* is **sorghum stem borer**.
356. *Culex fatigans* is responsible for transmitting **Filarial worms**.
357. Nematode develops cysts on potato? ***Globodera rostochiensis***.
358. Bubonic plague is a disease transmitted to man by **Rat Fleas**.
359. Seed gall nematode is ***Anguina tritici***.
360. Spiral nematode is ***Helicotylenchus***.
361. In neem, the insecticidal property is chiefly in the bioactive principle called **Azadirachtin**.
362. Powderpost beetles cause damage to **Wood**.
363. The common field rat that destroys crops in agricultural land is ***Bandicota bengalensis***.
364. The mouth parts of a book louse are of **Biting type**.
365. Inundative release is an activity associated with **Biological control**.
366. Insect thorax consists of **Three segments**.
367. Zinc phosphide is a **Rodenticide**.
368. The most commonly occurring root knot nematode in India is ***Meloidogyne incognita***.
369. The insect that can perceive U.V. light is ***Apis (honey bees)***.
370. In **Thysanura** Wings are absent.
371. The nematode associated with white tip disease in rice is ***Aphelenchoides besseyi***.
372. Ecdysone is chemically **Steroid**.
373. Vulval cone is present in ***Heterodera***.
374. **Protura** do not possess antennae.
375. **Metaldehyde** is a molluscicide.

376. **Fenvalerate** is a synthetic pyrethroid.
377. According to the 'International Code of Zoological Nomenclature', the ending "Oidea" is for **Superfamily**.
378. Egg-laying by bruchids on pulses is prevented by **Oil treatment**.
379. The insects whose adult life span is only for a few hours are **Ephemeropterans**.
380. EDCT is a mixture of ethylene dichloride and carbon tetrachloride in the ratio of **1:4**.
381. A pest that gains entry into the storage form field corp is ***Cyhlas formicarius***.
382. The origin of insects dates back to **350 million years**.
383. The polyphagous hairy caterpillar is ***Amsacta albistriga***.
384. A compound that is used to enhance the toxicity of an insecticide is **Synergist**.
385. The malaria parasite was first discovered by Sir Ronald Ross in **Hyderabad**.
386. Functional base of insect leg is **Coxa**.
387. Mouth parts in thrips belong to the **Asymmetrical type**.
388. The commercial formulation of *Bacillus thuringiensis* is **DIPEL**.
389. A serious pest of sorghum earheads is **Midge**.
390. Silk gland in silkworm is a modification of **Salivary gland**.
391. The vector of tomato spotted wilt virus causing bud necrosis in groundnut –**Thrips**.
392. Ber fruits are attacked by ***Carpomyia vesuviana***.
393. Name the serious pest of pulses in storage ***Callosobruchus maculates***.
394. *Bagrada hilaris* is a serious pest of **Mustard**.
395. *Aedes aegypti* is a vector of **Yellow fever**.
396. Entry of exotic pests can be checked through **Quarantine measures**.
397. *Pratylenchus* sp. are known as **Lesion nematodes**.
398. The human louse is ***Pediculus humanus***.
399. Chlorpyriphos is a **Organophosphate**.
400. The most prevalent domesticated silkworm in India is **Mulberry silkworm**.

401. *Trichogramma chilonis* is **Egg parasitoid of sugarcane borers.**
402. White earheads in paddy are due to **Rice stem borer**.
403. The lady bird beetle is a **Predator of aphids**.
404. Young leaves of citrus plants show serpentine galleries due to ***Phyllocnistis citrella***.
405. Plant parasitic nematodes generally attack **Roots**.
406. Virus transmitting nematodes are ***Xiphinema* and *Longidorus***.
407. Gall forming group of insects is **Cecidomyidae**.
408. Cerci are located on **Eleventh abdominal segment**.
409. Citrus nematode causes **Slow decline of citrus**.
410. Red ring disease of coconut is vectored by a weevil ***Rhynchophorus palmarum***.
411. Fore wings of Coleoptera are known as **Elytra**.
412. Which is an important nematode pest of coffee? ***Pratylenchus coffeae***.
413. The organization of nematode body is **Bilaterally symmetrical**.
414. Lac is produced by **A coccid**.
415. A common antidote for insecticide poisoning is **Atropine**.
416. The number of chambers in the heart of cockroach is **13**.
417. The leaf miners belong to the family **Gelichiidae and Agromyzidae**.
418. The pupa of Hymenoptera are **Exarate**.
419. The most serious sucking pest of cotton is **Cotton leaf hopper**.
420. Summer diapause is known as **Aestivation**.
421. Extremely mammalian toxic pesticides label is **Bright red**.
422. Nodus is present in the wings of **Odonata**.
423. Elephantiasis is caused by ***Wuchereria bancrofti***.
424. Which nematodes does not produce eggs in gelatinous matrix? ***Meloidogyne***.
425. *Musca domestica* lays eggs in **Cow dung**.
426. What is an example of entomophilic nematode? ***Heterorhabditis indicus***.
427. The major component of insect cuticle is **Chitin**.
428. Which one forms 'nematode wool'? ***Ditylenchus***.
429. Pest which occurs in isolated locations is known as **Sporadic pest**.
430. Study of animal behaviour is known as **Ethology**.

431. The largest insect is **Atlas moth.**
432. A pest of quarantine importance in mango is **Mango nut weevil.**
433. Giant cells are produced by ***Meloidogyne***.
434. In root-knot and cyst forming nematodes, the infective stage is **Young female.**
435. A serious pest of bee hives is ***Galleria***.
436. If a pest is confined to a particular area and occurs regularly. Its infestation is termed as **Endemic.**
437. Malathion is an / a **Organophospate insecticide.**
438. Fore wing of coleoptera are known as **Elytra.**
439. Probosis in butterfly is formed by **Galea.**
440. The hearing organ in insects is Johnston's organ Absorption of digestion products in insects mostly occur in **Hind gut.**
441. Muga silkworm *Antheraea assamia* belongs to the family **Saturnidae.**
442. Scent producing scales on the wings of male moth are known as **Androconia.**
443. The scientific name of tea mosquito bug is ***Helopetis theivora***.
444. Study of animal behavior is know as **Ethology**.
445. Silver fish belongs to the order **Thysanura.**
446. A secondary storage pest is **Flat grain beetle.**
447. Biological control agent of agricultural pests? **Reduvidae.**
448. The vector of dengue fever is ***Aedes eagyptii***.
449. Bunchy top in sugarcane is due to **Top shoot borer.**
450. Kala-azar is transmitted by **Sand fly.**
451. The scientific name of mustard aphid is ***Lipaphis erysimi***.
452. The tetranychid mite has the following number of nymphal stages **3.**
453. A serious pest of pulses in storage is ***Callosobruchus chinensis***.
454. Human onchoceresis is caused by **Nematode.**
455. Pterostigma is a distinguishing feature of **Odonota.**
456. A pest of quarantine importance in mango is **Mango nut weevil.**
457. An introduced pest ***Plutella xylostella***.
458. Mosquito mat contains **Allethrin.**
459. The scientific mane of the rock bee is ***Apis dorsata***.
460. Pebrine disease in silkworm is caused by **Protozoa.**

461. Tukra disease in mulberry is caused by **Mealy bugs.**
462. The ootheca of *Periplanta american* contains **14-16 eggs.**
463. Ear cockle nematode is a pest of **Wheat.**
464. Silk contains **Sericin and Fibroin.**
465. The nematode associated with panama Wilt of banana is **Burrowing nematode.**
466. Silk glands are modified **Labial glands.**
467. A serious pest of bee hive is ***Galleria mellonella*.**
468. The mouth spear in nematode is nothing but **Stylet.**
469. Lac cultivation is mostly practiced in **Bihar.**
470. Perennial pattern is important for identification of ***Meloidogyne* sp.**
471. Parasite of coconut black headed caterpillar is ***Tricospilus pupivora*.**
472. Inundative release is an activity associated with **Biological control.**
473. Green muscardine fungi is ***Metarrhizium anisoplae*.**
474. The causal organism for molya disease is ***Heteroder avenae*.**
475. An example of moluscicide is **Metaldehyde.**
476. An example of systemic insecticide is **Monocrotophos.**
477. An example of pest resurgence due to use of pyrethroid is **Whitefly.**
478. Insecticide resistance is observed in **Diamond back moth.**
479. The common sprayer used for spraying fruit trees is **Rocker sprayer.**
480. Insect blood cell? **Plasmatocyte.**
481. Hemoglobin is present in **Chironomid larvae.**
482. Cerci are paired appendages present **At the hind end of an insects abdomen.**
483. The presence of filter chamber is unique of **Homoptera.**
484. Damage caused to paddy by *Leptocorisa varicornis* produces **A hole where a white or brown spot develops and drying up of ears.**
485. *Leucinodes orbonalis* infests vegetables? **Egg plant.**
486. The colour of eggs of *Erias vittella* is **Bluish green.**
487. *Bagrada picta* infests **Mustard.**
488. The secondary host of *Leishmania donovani* is ***Phlebotomus*.**
489. One of the distinguish feature of dermestid beetle is **Larval body covered with hairs of various length.**

490. Members of the genus *Componotus* are known as **Carpenter ants.**
491. Appearance of small holes and piles of saw dust on the wooden materials are symptoms of infestation by **Powder post beetles.**
492. On encountering an unexpected obstacle book lice normally **Retreat rapidly backwards for a short distance.**
493. 'Ufra' disease of rice is caused by ***Ditylenchous angustus***.
494. Rangeeni is a type of **Lack strain.**
495. Dammar bee is zoologically known as ***Melipona iridipennis***.
496. An insect involved in the biocontrol of *Parthenium* **Zygogramma bicolarata.**
497. Inhibition of cholinesterase enzyme is brought by **Organophosphorus and carbamates.**
498. Eggs of *Corcyra* are used as a medium for culturing the biocontrol organisms ***Trichogramma* sp.**
499. Chitin inhibitor? **Dimilin.**
500. *Epiricania melanoleuca* is an insect biocontrol agent belongs to the order **Lepidoptera.**
501. Use of irradiation as a physical control measure in insect suppression programme maximally affects **Larval stage.**
502. Nodus is present in wings of **Odonata.**
503. Peritropic membrane is the inner most layer of **Mesentron.**
504. In insect wings the coastal margin is located **Along anterior margin.**
505. The main pulsating organ of circulatory system lies in **Head, base of wings, thorax and antennae.**
506. In females Spermatozoa are stored in **Spermatheca.**
507. Antennae are innervated by **Deutocerebrum.**
508. In apneustic respiratory system **No spiracle is functional.**
509. Silvery shoot in onion is caused by ***Thrips tabaci***.
510. *Idioscopus clypealis* is a serious pest of **Mango.**
511. The most common fumigant for bulk storage of wheat is **Aluminium phosphide.**
512. *Holotrichia consanguinia* is the most serious pest of **Ground nut.**
513. *Bovicola caprae* is the common louse of **Goats.**
514. The mange disease in dogs is caused by ***Demodex canis***.
515. *Calliphora erythrocephala* is commonly attracted to **Meat.**

516. Bed bugs belongs to the order **Hemiptera.**
517. The unpleasant odour in food articles is imparted by the body secretions of **Cockroaches.**
518. In kitchen turmeric is usually attacked by ***Lasioderma serricorne*.**
519. White tips in paddy leaves are caused by ***Aphelenchoides beseeyi*.**
520. Nemagon acts as a good **Nematicide.**
521. The common field rat that destroys crops is ***Bandicota bengalensis*.**
522. Rational use of insecticide is possible through the adaptation of **Economic threshold level.**
523. In honey bees propodeum is **Fusion of first abdominal segments with metathorax.**
524. Trichogen cell gives rise to **Setae.**
525. Head includes parts of **6.**
526. Citrus canker transmitted by **leafminer.**
527. Dipel, Halt is ***Bt*.**
528. Mites bellow coconut sap, belong to family **Eriophyidae.**
529. *Aceria cajani* is a mite vector of **Pigeonpea sterility mosaic.**
530. Radio active dating method used to identify **Fossils.**
531. First resistance noticed in **Singara beetle.**
532. Bordeux mixture is composed of **Copper sulphate and Quick lime.**
533. An important object of the ICZN is to provide **Uniqueness, Stability, Distinctiveness to the scientific names.**
534. Until 2002, the most recent insect order was Grylloblattodea described in 1914. But in 2002, a new order was described with the name **Mantophasmatodea.**
535. Storage excretion is the function of **Urate cells.**
536. Sub-social insects living in silken tunnels are **Embiids.**
537. The larva construct pit fall traps in **Ant lions.**
538. Phase polymorphism is shown by **locust.**
539. Plastron respiration is observed in **Aquatic insects.**
540. predominant excretory end product of protein breaks down in terrestrial insect is **Uric acid.**
541. *Cryptolaemus* sp. is mass multiplied in the laboratory on **Mealy bugs.**

542. The weed killer for the suppression of *Chromolaena odorata* is ***Procedochares***.
543. *Chrysoperla* sp. is mass multiplied on the **Egg stage of rice moth**.
544. An insecticide safer to honey bees is **Phosalone**.
545. An extremely toxic insecticide will have LD 50 of **1-50 mg/kg**.
546. A toxin isolated from marine annelid is **Nereistoxin**.
547. Methyl eugenol is a food attractant for **Oriental fruit fly**.
548. The major detoxifying enzymes involved in resistance to pyrethroids are **MFO's**.
549. Avermectins are derived from ***Streptomycis* sp.**
550. The predominant sugar in insect haemolymph is **Trehalose**.
551. The vitellogenesis synthesis in fat body is stimulated by **JH**.
552. Drooping of coffee berries with a hole in the naval region is due to ***Hypothenemus hampei***.
553. Bunchy top in sugarcane is caused by ***Scirpophaga* sp.**
554. A *B. t.* subspecies that are specific to maggots and caterpillars are ***isralensis* and *kurstaki***.
555. Navigation in honey bees is facilitated through: **Polarized light**.
556. Thai sac brood disease has been devasting on ***Apis mellifera*** bee in India.
557. Larval parasitoid affecting early instar of *Pectinophora gossypiella* ***Apanteles angulati***.
558. During insect locomotion the body always rests on **Tripod**.
559. Hermetic storage means **Storage excluding atmospheric O_2 and permitting CO_2 accumulation**.
560. A liquid fumigant? **Methyl bromide**.
561. Founder editor of the Journal "Pesticide Biochemistry and Physiology" is **Richard O'Brien**.
562. No. of stylets in Mallophaga are **Three**.
563. Geomagnetic receptors are well developed in **Termite**.
564. Co-ordinating centre for feeding is **Sub-esophagial ganglion**.
565. Saber like ovipositor is seen in **Tenthredinidae**.
566. Barklice belong to the family (order) is **Posocoptera**.
567. Malaise trap is useful in trapping **Flying insects**.
568. When market value increases **EIL decreases**.

569. Ryania, the muscle poison inhibits **Ca^{++} channel.**
570. Boric acid is a **Physical poison.**
571. In which case Bt response is involved **Antibiosis.**
572. Hopper burn in rice is due to the feeding of ***Nilaparvata lugens***.
573. A reduvid predaceous on red cotton bug is ***Harpactor costalis***.
574. Physapod refers to the order **Thysanoptera.**
575. Mutagenecity test for carcinogenic potential of chemicals is **Ames test.**
576. Non-reelable silk is obtained from ***P. ricina***.
577. Fossorial leg present in **Gryllotalpidae.**
578. Quick method to determine hidden infestation in stored product is **Acid fusclin.**
579. A tank is absent in **Foot sprayer.**
580. Hardness of elytra is due to **Presence of trabeculae.**
581. In honeybees, Front legs are modified for **antennae cleaning.**
582. Predator of honey bee ***Merops orientalis***.
583. Major sugars found in honey are **Glucose + Fructose.**
584. The honey bee species that does not sting ***Mellipona irridipennis***.
585. Bee venom mainly consists of **Mellitin.**
586. Queen cage is used for **Preventing swarming.**
587. Protein source for honey bee is **Pollen.**
588. Wing coupling mechanism in honey bee is **Hamuli type.**
589. Honey bee does not perceive **Red light.**
590. Middle pair of legs in honey bee is useful for **Pollen depositing in cells and wax picking.**
591. The major nitrogenous polysaccharide present in insect cuticle is **Chitin.**
592. Johanston's organ is located in **Pedicel of antenna.**
593. The second segment of an insect leg is **Trochanter.**
594. The sensilla that perceive odours are **Olfactory sensilla.**
595. In insects, the major part of digestion and absorption of food takes place in **Midgut.**
596. The major function of filter chamber is to remove rapidly the excess **Water.**
597. The Malpighian tubules are associated with **Excretion.**

598. The larvae of moth are **Polypod type.**
599. A pair of tubular structures found on the dorsum of aphid abdomen are **Cornicles.**
600. In Coleoptera, the forewings are modified into **Elytra.**
601. The cochineal insects belong to the taxonomic group **Dactylopiidae.**
602. The hawk moths belong to the family **Sphingidae.**
603. The wings of moths or butterflies are covered with **Scales.**
604. The family that includes exclusive egg parasitoids is **Trichogrammatidae.**
605. The fruit flies belong to the family **Tephritidae.**
606. The insect viruses are recommended in unit of **LE/ha.**
607. An example for systemic foliar insecticide is **Methyl demeton.**
608. The major terpenoid present in the neem seed is **Azadirachtin.**
609. The trap crop for managing tomato fruit borer is **Marigold.**
610. Trimming of bunds in paddy fields helps ins suppression of **Grass hopper.**
611. Preserve of silken tubular galleries in stored products is an indication of presence of **Rice moth.**
612. An insecticide with nematicidal action **Carbofuran.**
613. An example for acaricide **Dicofol.**
614. The insecticidal property of DDT was discovered by Paul Muller in the year **1939.**
615. Thrips have **Rasping and Sucking** type of mouth parts.
616. The layer of insect cuticle responsible for water conservation is **Wax layer.**
617. The dorsal sclerite of insect body segment is **Notum.**
618. The first pair of insect wings are borne on **Mesothoracic segment.**
619. Hamulate type of wing coupling apparatus is seen in **Bees.**
620. In insect midgut, the **columnar cells** are involved in secretion of digestive enzyme.
621. The membrane that protects midgut epithelium in insect is **Peritrophic.**
622. The principal pulsatory organ for blood circulation in insect is **Dorsal blood vessel.**
623. The larvae of flies are **Apodous type.**

624. The clinging type of leg modification is seen in **Sucking louse.**
625. Whiteflies belong to the taxonomic group **Aleyrodidae.**
626. The snout moths belong to the family **Pyralidae.**
627. Thrips have **Fringed type of wings.**
628. The weevils belong to the taxonomic group **Curculionidae.**
629. An example for systemic soil insecticide is **Phorate.**
630. Draining out of water for a few days in paddy field helps in suppression of **BPH.**
631. The insecticide used for stem injection against *Opisina arenosella* **Monocrotophos.**
632. Among the insect pathogens, the most promising biocontrol agents are **Bacteria.**
633. The moisture content of food grains for safe storage should be less than 9 per cent.
634. Hairy caterpillars belong to the family **Arctiidae.**
635. The insecticide derived from an actinomycetes organism **Avermectin.**
636. The snout moths belong to the family **Pyralidae.**
637. A pathogenic fungus for control of coconut mite ***Hirsutella* sp**.
638. *Bacillus thuringiensis kurstaki* is specific to **Caterpillars.**
639. Contraction of heart during its beat is called **Systole.**
640. Young ones of dragon flies are **Naiads.**
641. Directorate of plant protection and storage is present at **Faridabad.**
642. Chitin is **Nitrogenous polysaccharide.**
643. In winged insects meso and meta thorax are together called as **Pterothorax.**
644. Grouping of segments in to body regions is called **Tagmosis.**
645. The facets of a compound eye are called **Ommatidia.**
646. Insects undergo resting due to low temperature in winter is called **Hibernation.**
647. In honey bees pollen basket and pollen brush are seen in **Hind legs.**
648. When immature insects reproduce it is called **Paedogenesis.**
649. Insect blood does not contain **Vitamin K.**
650. Stemmata are the eyes present in **Caterpillars.**
651. Cantharidin is a product from insects of the family **Meloidae.**
652. Boll worms are pests of **Cotton.**

653. Proboscis of butterflies is a modified **Galea.**
654. Honey bee sting is modified **Ovipositor.**
655. Insect systematic is the study of **Phylogeny and systematic.**
656. The pupae of mosquitoes are called **Tumblers.**
657. Skin winged insects are **Dermoptera.**
658. The function of spermatheca in insects is to **Store sperms.**
659. Spiracles mostly present on **Pleurites.**
660. Moulting hormone is produced by **Prothoracic glands.**
661. Number of stylates in piercing and sucking types of mouth parts of bugs and female mosquitoes respectively are **4, 6.**
662. Bhopal gas tragedy caused world wide drop in the production of insecticide **Methyl isocyanite.**
663. The most economic and convenient prophylactic control measure is **Cultural control.**
664. The nerve cells that transmits the stimuli from ganglia to effectors organs are called **Motor nerves.**
665. Pore canal is found in the layer of **Exocuticle.**
666. Anticoagulants are secreted by **Mosquitoes.**
667. A pair of upper jaws present in mouth parts these are called as **Mandibles.**
668. Adults live for few hours in **Ephemeroptera.**
669. Among these layers of body wall which is the cellular layer **Epidermis.**
670. Amplexiform type of wing coupling mechanism is seen in **Butterflies.**
671. Anopheles mosquito eggs are **singly flaoting.**
672. Aristate type of Antennae is present in **House fly.**
673. Below clypeus we find **Labrum.**
674. Bifid claw present in **Pieridae.**
675. Body processes/appendages which are incapable of movement **Spines.**
676. Butterfly pupae are called as **Chrysalis.**
677. Carbiculum is modified for **Pollen basket.**
678. Carinae is present in **Lophopidae.**
679. Cement layer is the part of **Epicuticle.**

680. Cockroaches have type of antennae called **Setaceous type.**
681. Compound eyes are consist of smaller eyes called as **Ommatidia.**
682. Compound eyes which are separate each other, condition is called as **Dioptic eyes.**
683. Corbiculate type of legs are present in **Worker honey bees.**
684. Corium is present in **Egg.**
685. Elytron type of wings present in **Beetles.**
686. **Excretory system** is responsible for Homeostatic.
687. Eyes which are involve in only regulation of light intensity **Simple eyes.**
688. Fenestrae present in **Cockroach.**
689. First tarsal segment is called as **Basi tarsus.**
690. For construction of comb, honey bees uses **Mandibles.**
691. For wing articulation and for musule attachment the protein involved is **Resilin.**
692. Fore wings of cockroach are called as **Tegmina.**
693. Forewings are present in **Mesothorax.**
694. Fossorial type of legs means **Digging type.**
695. Frequency of wing beat is expressed in terms of **Hz/ sec.**
696. Fringed wings present in **Thrips.**
697. Gustatory receptors are associated with **taste.**
698. Hemimetabolous development is also called as **Indirect metamorphosis.**
699. Hyper metamorphosis is seen in **Strepsiptera.**
700. Hyper pneumatic type of tracheal system is found in **Diplura.**
701. In Antennae cleaner type of legs the modified parts are **Tibia and Basitarsus.**
702. In cicadidae, tympanal organs are present in **1st abdominal segment.**
703. In embioptera silk glands are present on **Basitarsus.**
704. In fireflies, light is produced due to oxidation of **Luciferin.**
705. In grass hopper which legs are modified for jumping purpose **Hind legs.**
706. In honey bees the meta thorax is fused with the I abdominal segment called **Propodium.**

707. In Hover flies spurious vein present between **R & M.**
708. In male cicada, tymbal is present in **1st abdominal segment.**
709. In Nepidae hind legs modified for **Swimming.**
710. In pollen carrying type the modified parts of legs are **Tibia and Basitarsus.**
711. In strepsiptera Fore wings are **reduced.**
712. In thrips, asymmetrical type of mouth part is because of **Right mandibles are absent.**
713. Insect head consist of **6 segments.**
714. Insects which do not undergo metamorphosis are come under the group **Ametabola.**
715. Insects which jump without using any legs **Maggots of flies.**
716. Johnston's organ present in **Pedicel.**
717. Johnston's organs are present in **Pedicel** part of the antenna.
718. Jugate and frenate type of wing coupling mechanism is seen in **Moths.**
719. Last part / segment of leg is called **Pre tarsus.**
720. Lateral plate of thorax is called as **Pleuron.**
721. Lateral side of head is called **Gena.**
722. Leg consist of primitively **6 segments.**
723. Male genitalia in odonata present on **2 and 3 abdominal sterna.**
724. Male Mosquitoes contain **Plumose type of** antennae.
725. Malpighian tubules absent in **Aphids and Collembola.**
726. Moulting fluid during moulting is secreted by **Epidermis.**
727. Moulting fluid is produce from **Epidermis.**
728. Moulting fluid is produced by **Hypodermis layer** of the integument.
729. Naiads of Anisoptera have **Rectal gills.**
730. Nitrogenous polysacharide present in insect integument is **Chitin.**
731. Non chitinous layer of integument in insects is **Epicuticle and Endocuticle.**
732. Non-chitin as part of the cuticle is **Epicuticle.**
733. Number of recurrent veins present in Ichneumohidae **2.**
734. Ocelli in cockroach are called as **Fenestrae.**
735. Outer layer of egg is called **Chorion.**
736. Outermost layer of body wall is **Cement layer.**

737. Pheromones are used for **Intra specific communication.**
738. Plume moth have **fissured wings.**
739. Pore canals are present in which layer of the insect integument **Epicuticle.**
740. Primitive or basic type of mouth part is **Biting and chewing.**
741. Ptero thorax made up of **Meso and Meta thorax.**
742. Puparium is nothing but **Co-arctate.**
743. Rasping and sucking type of mouth parts present in **Thrips.**
744. Roya jelly is secreted by **Hypopharyngeal glands.**
745. Segmental plate which is present on dorsal side of thorax is called as **Tergum / Notum.**
746. Semicircular hind wings are seen in **Earwigs.**
747. Similar form of wings are fond in **Isoptera.**
748. Some immature insect are able to reproduce is called **Paedogenesis.**
749. Some insects do not have malpighian tubules like **Collembolans and Aphids.**
750. Technical term of head capsule is **Cranium.**
751. The layer of Epicuticle, which prevents water loss is **Lipid layer**.
752. The layer which separates epidermis from the haemocoel is **Basement membrane.**
753. The layer which serves as a water proof layer of the integument is **Wax layer.**
754. The line of weakness is present in /on Head region called as **Epicranial suture.**
755. The only cellular layer of integument is **Epidermis.**
756. The protein which is involved in hardening of cuticle **Sclerotin.**
757. The shape of eggs of Bugs is **Barrel**.
758. The thickest part of the leg **Femur**.
759. Thickest layer of the integument is **Exocuticle.**
760. Thorax is made up of **3 segments.**
761. Typical antennae contain **3 parts.**
762. Wide spread excitatory transmitter is **Acetylcholine.**
763. Wing flexion mechanism well developed in **Neoptera**.
764. A median muscular pad present in between a pair of lateral claws in the foot of insects is called as **Arolium.**

765. Wooly aphid is the serious pest of **Sugarcane.**
766. The first *Bt* transgenic crop under field testing in India is **Cotton.**
767. Which is absent in arthropoda? **Closed blood vascular system.**
768. Prickly pear is eradicated by the insect ***Dactylopius tomentosus***.
769. Hopper burn symptom is due to **Leaf hopper.**
770. The stem borer which is common to sorghum, maize, rice and wheat is ***Sesamia inferens***.
771. Curling of leaves in chillies is caused due to **Thrips.**
772. Morphologically identical but reproductively isolated species **sibling species.**
773. Blister beetles undergo **hypermetamorphosis.**
774. Parasite of a parasite is **Hyperparasite.**
775. *Coccinella septumpuctata* is **predator.**
776. Unique character of Arthropods is having **segmented appendages.**
777. Lamellar structure to the insect cuticle is due to **microfibres.**
778. Tormogen cell is associated with **seta.**
779. Collembola conserve cuticular materials by **having very little exocuticle and eating their own exuviae.**
780. Presence of membrane between the sclerites gives **flexibility**.
781. Ptilinum is found in the larvae of **Diptera.**
782. Acrotergite is associated with **tergum.**
783. Coronal suture is part of **epicranial suture.**
784. Hydrofuge properties of insect cuticle is largely due to **density of wax layer.**
785. Antecostal suture is located on **tergum.**
786. Most important feature of insects is having **three pairs of legs.**
787. Biramous appendages are found in **Crustacea.**
788. Trichoid sensilla is associated with **tactile.**
789. Cervical Sclerites are associated with **cervix.**
790. Median plates are associated with **M & Cu veins.**
791. Trochantin is usually occurs in **Coleoptera.**
792. Pecten is associated with hind leg of **honey bee.**
793. Claval suture is associated with **wing.**
794. Transportation of cuticular substances to the epicuticle is done by **pore canals.**

795. Bleaching agent to clear the wings of lepidopterans is mixture of **Sodium chloride and sodium hypochlorite.**
796. Procephalon and gnathocephalon are fused to form **head.**
797. Insects have originated about **400 million years back.**
798. Aeropyles are concerned with **respiration.**
799. Honey bee sting is a modified **ovipositor.**
800. Median caudal filament is present in **thysanura.**
801. Single large claw is present in **clinging.**
802. The alimentary canal of insect is derived from **Ectoderm and Endoderm.**
803. The nervous system of insect is derived from **Ectoderm.**
804. Moulting fluid is able to digest **Endocuticle.**
805. The process of post ecdysial hardening and darkening of the cuticle is regulated by **Bursicon.**
806. Glycogen is **Polysacchride.**
807. Which of the following enzyme is not associated with the digestion of food **Phenoloxidase.**
808. The PTTH is secreted by **Neurosecretory cells.**
809. Most of the insects excrete 80 to 90 percent of their nitrogen waste in the form of **Uric acid.**
810. In which organelle protein sysnthesis takes place **Ribosome.**
811. Ketone bodies are oxidized by **Flightmuscles.**
812. Oxidation of fattyacids takes place into **Ribosome.**
813. In Glycolysis 10 molecules of ATP are produced but the net gain is **8 molecules.**
814. Glycerophosphate shuttle is found in **Flight muscles.**
815. Which of the following enters the Krebs cycle **Acetyl COA.**
816. Which of the following are not haemocytes **Nephrocytes.**
817. The gland which produces pheromone in queen honeybee **Mandibular gland.**
818. Which of the following organs are not involved in excretion **Hepatic caeca.**
819. Trehalose is a **Trisacchride.**
820. Aminoacid that cannot be senthesised in the insect body **Methionine**
821. 1 ATP is equivalent to **7.6 Kcal.**

822. The cytoplasm of the muscle fibre is called **Sarcoplasm.**
823. The process of oocyte leaving the ovariole is called **Ovulation**.
824. Pest management intervention must be applied at **ETL.**
825. NCIPM was 1st established in **1988.**
826. The concept of integrated pest management emphasizes an approach which is **Ecologically compatible.**
827. C:B ratio can be computed by **Net profit.**
828. At ETL the C:B ratio is **1:1.**
829. For management of whitegrub in sugarcane the best time for application of *M. anisopliae* with FYM is **May- June.**
830. Example for ingesting microbes are **Protozoa, virus, *Bt*.**
831. All biopesticides are registered under CIB guidelines under **9(3B)** rule.
832. Among the natural enemies used in biocontrol order of preference should be **Parasitoid- Predator- Pathogen.**
833. Most commonly used laboratory host for multiplication of parasitoids, predators and nematode is ***Corcyra cephalonica*.**
834. One CC of *Corcyra cephalonica* contains **18,000-20,000** number of eggs.
835. To get higher sporulation of entomopathogenic fungi commonly used media is **SMAY media.**
836. 'White halo' is the common name for which of the following pathogen ***Lecanicillium lecanii.***
837. 'Tree top' disease symptom is produced by **NPV.**
838. The first journal on biological control, Entomophaga has changed its name in 1997 to **Journal of biological control.**
839. Most practical biological control scientist is **Koebele.**
840. One of the corner stone of IPM programme is **Biological method.**
841. *Cyrtobhagus salviniae* is an effective weed control agent against **Water fern.**
842. More than 50% of the successful BC agents in Hymenopteran parasitoides belongs to **Encyrtidae, Aphelinidae.**
843. The term 'parasitoid' was proposed for insect parasites by **Reuter**.
844. 50 per cent of the predators belongs to **Coleoptera.**

845. *Eichornia crassipes* is effectively controlled by ***Neochetina bruchi, Neochetina eicchorniae*** **and** ***Orthagalumna terebrantis***.

846. The effective parasitoid on papaya mealybug is ***Acerophagus papayae***.

847. Monocrotophos resistant predator developed is ***Chrysoperla carnea***.

848. Thermophilic resistant parasitoid is ***Trichogramma chilonis***.

849. Example for 'homoxenous parasitoid' is ***Platygaster oryzae***.

850. *Isotima javensis* is a **larval-pupal parasitoid.**

851. European corn borer undergoes diapause in **stalks, stubbles & cobs.**

852. How many generations has Deccan wingless grasshopper in a year? **one.**

853. Which pest/s minimize due to clipping of leaf tips before transplanting in paddy? **stem borer and thrips.**

854. Which is the destructive stage of shoot fly in sorghum? **maggot.**

855. DSV-3, a sorghum variety resistant to which pest? **earhead midge.**

856. Trap crop used for management of *Spodoptera litura* in Groundnut **castor.**

857. Bud necrosis is caused due to ***Thrips palmi***.

858. Parasitoid used to control seeding borer in sugarcane ***Trichogramma chilonis***.

859. Red headed hairy caterpillar of groundnut can be effectively managed by **Light traps.**

860. *Anomis sabulifera* is a serious pest in which crops **mestha.**

861. *Ceratoracana lanigera* is most destructive to which crop **sugarcane.**

862. Which crop is attacked by *Tetraneura nigriadominalis* **ragi.**

863. Alternate host of groundnut leaf miner **soybean.**

864. *Atherigona naqeil* is **Polyphagous**.

865. Floral pest of linseed ***Dasyneura lini***.

866. Which plant part/s is attacked by *Dichlocrocis punctiferalis* **shoot and capsules.**

867. Flower bud maggot or gall midge of cotton ***Dasineura gossypii***.

868. Effective insecticide for mired bugs control in cotton **Acephate.**

869. Which sugarcane pest is predated by *Micromus igoratus*? **woolly aphid.**

870. Economic threshold level for *Helicoverpa armigera* in cotton **one larva/plant.**
871. Trap crop used in cotton for shoot weevil **okra.**
872. The term Entomology has been derived from: **Greek word.**
873. Carbofuran 3G belongs to: **Carbamate.**
874. *Eublemma amabilis* is an important enemy of: **Lac insect.**
875. DDT belongs to: **Organochlorines.**
876. Concentrate liquids are used for: **Ultra low volume spraying.**
877. The scientific name of Indian honey bee is: ***Apis cerana.***
878. Termites have type of antennae: **Moniliform.**
879. An insecticide with short residual action: **Malathion.**
880. Third segment of the leg: **Femur.**
881. Pollu beetle is a major pest of: **Cardamom.**
882. Colour of chironomid larval blood is: **Red.**
883. Arthropoda comprises of: **90% of the known species in the animal kingdom.**
884. Mulberry variety best suited for young age (chawki) silkworm is: **S-36.**
885. Inundative release is a methodology associated with: **Biological control.**
886. Apodus larvae have: **No legs.**
887. Damaging group of termite is: **Workers.**
888. A blood sucking fly attacking cattle: **Horse fly.**
889. Mandibulosuctorial type of mouth parts is found in: **Grub of antlion.**
890. Trenching is one of the following management: **Mechanical.**
891. Methyl eugenol is used against: **Fruit fly.**
892. Which trap crops is used for the control of *Spodoptera litura*: **Castor.**
893. Example of Bt transgenic cotton: **Kanaka.**
894. Johnston's organ is present in: **Pedicel.**
895. Gills help naids to: **Respire.**
896. *Bacillus thuringiensis* is a: **Bacteria.**
897. *Paratelphusa hydrodromus* is mostly found in: **Rice field.**

898. *Nomuraea rileyi* is: **Fungus.**
899. Termites belong to the order: **Isoptera.**
900. Little leaf of brinjal is caused by a: **Phytoplasma.**
901. The trade name of monocrotophos is: **Nuvacron.**
902. Basic unit of classification is: **Species.**
903. Insecticide generally formulated as granule: **Carbofuran.**
904. Preying mantids have legs adapted for: **For catching small insects (raptorial).**
905. Ligula is part of: **Labium.**
906. Diapause in *Amsacta albistrigra* is seen in: **Pupal stage.**
907. Cephalothorax is the term given to body segment of: **prawn.**
908. Rasping and sucking type of mouth part is: **Thrips.**
909. Entomology is: **Study of insects.**
910. Grass hopper's hind legs are adopted for: **Jumping.**
911. Metaldehyde is recommended for the control of: **Molluscs.**
912. The time taken to kill 50 per cent population of the test insect is: **LT_{50}.**
913. Only prothoracic spiracles are functional in case of: **Propneustic.**
914. In *Bombyx mori* diapause occurs at: **Egg stage.**
915. *Micromus igorotus* an efficient predator of sugarcane woolly aphid belongs to: **Neuroptera.**
916. The Govt. of India enacted the Central Insecticide Act in the year **1971.**
917. Housefly belongs to the family: **Muscidae.**
918. Tympanum is found in lateral side of first abdominal segment in: **Short horned grasshopper.**
919. Vasiform orifice is found in the members of the: **Aleyrodidae.**
920. Endocrine gland produces: **Juvenile hormones.**
921. Plastron respiration occurs in: **Aquatic insects.**
922. In nature if ER> BP there will be: **Decrease in insect population.**
923. First successful biological control of an insect was achieved by using: ***Rodalia cardinalis*.**
924. Neuropteran predator is ***Chrysoperla*.**
925. Only one pair of wings are found in: **Diptera.**

926. Eggs of head louse are called as: **nits.**
927. In honey bees, the queen substance is secreted by: **Mandibular glands.**
928. Presence of bored holes on tomato is due to: ***Helicoverpa*** **and** ***Spodoptera*****.**
929. Binomial nomenclature is used to name: **Species.**
930. Potassium cyanide is a: **Protoplasmic poison.**
931. Sub family name always ends with: **inae.**
932. *Bacillus thuringiensis* is most effective at pH: **9 to 10.**
933. Chaffyness of individual grains in paddy earheads is because of: **Earhead bug.**
934. The tongue of insects is also called: **Hypopharynx.**
935. Bee dance communicates: **Distance to the food source, Direction of food source and Richness of food source.**
936. Chemicals which are used to kill mites are called **Acaricides.**
937. Ladybird beetle is predator of: **Cotton cushion scale.**
938. Endosulfan is an example for: **Organo chlorine group of insecticides.**
939. Ferrolure is used in the management of: **Red palm weevil.**
940. Insect head is made up of: **6 segments.**
941. Sericin is: **Silk protein.**
942. Mexican beetle is an example for: **Weed killer.**
943. An example for systemic insecticide: **Dimethoate.**
944. Type of leg present in housefly: **Climbing.**
945. Dead heart in crops is a symptom due to attack of: **Stem borers.**
946. Hydrofuge properties of insect cuticle is largely due to: **Wax mono layer.**
947. Two pairs of appendages per segment are present in: **Millipedes.**
948. Lamellar structure to the insect cuticle is due to: **Microfibres.**
949. Oenocytes are concerned with: **Moulting cycle, Secretion of lipoprotein of epicuticle and Synthesis of wax.**
950. Molecules of wax mono layer are inclined at an angle of: **25 degrees.**
951. Opisthognathous head is found in: Leafhoppers, **Mealy bugs and Scale insects.**

952. Proteinase and chitinase digest - **Endocuticle.**
953. Antecostal suture is located on - **Tergum.**
954. Which sampling strategy is best suited for very small insects that live in the soil? **Berlese funnel.**
955. Study of animal behaviour is called as - **Ethology.**
956. The insecticide with the isomer HCH (BHC) used now in pest control is - **Lindane.**
957. The presence of filter chamber is unique feature of - **Homoptera.**
958. Black drongo a potential predator on caterpillar belongs to - **Bird.**
959. *Bacillus thuriengensis* Var. thuriegnensis is used against - **Caterpillars.**
960. The fungus popularly called as "White halo" - ***Verticillum lecani.***
961. The single largest group of insecticides of total registered pesticide in the world is - **Organophosphates.**
962. NBAIR is located at - **Bangalore.**
963. The organ that helps in excretion and regulation of internal environment in insects by controlling ion and water balance in blood is - **Malpighian tubules.**
964. Dengue disease is transmitted by ***Aedes aegypii.***
965. Worker bee is **Sterile female.**
966. Mosquito coil contains **Allethrin.**
967. Biotic potential is a result of **Fecundity, Sex ratio and Mating potential.**
968. A pest which has a single host is referred as **Monophagous.**
969. *Galleria Mellnella* is pest of **Bees.**
970. Insects belongs to the phylum **Arthopoda.**
971. Honey bees have mouth parts of the type **Chewing and Lapping.**
972. Plant bugs belongs to the order **Hemiptera.**
973. Brunchids are serious pests on **Pulses.**
974. Ants have the type of antennae: **Geniculate type.**
975. Forewings are present on **Meso thorax.**
976. *Zygogramma bicolorata* is a **Weed killer.**
977. Thrips belongs to the order **Thysanoptera.**
978. A parasite on a parasite is called **Hyper parasite.**
979. Methyl bromide is used against **Storage pests.**

980. Thai sac brood is a severe disease of **Bees.**
981. *Metarhizium anisopliae* is an entomopathogenic **Fungi.**
982. Bt cotton is resistant to **Bollworms.**
983. "Tree top disease" symptoms noticed in lepidopteran larvae when they are infected with **NPV.**
984. Four insecticide containers have following colour markings on their container, which colour indicates most dangerous to human beings apart from insects **Red.**
985. *Goniozus nephantidis* is a parasitoid of **CBHC.**
986. The major pests of bee comb is **Wax moth.**
987. The eco-friendly insecticide is **Nimbicidine.**
988. Prickly pear weed has been effectively controlled by ***Dactylopius tomentosus.***
989. Crop acts as a trap crop for cabbage DBM **Mustard.**
990. Host used in *Trichogramma* mass production is **Rice moth.**
991. The insecticidal property of DDT was discovered by **Paul Muller.**
992. Poison bait is very effective against **Sorghum armyworm.**
993. Imidacloprid 17.8 SL is available in the market under the trade name as **Confidor.**
994. Occurrence of pest on a regular basis at a particular place **Persistant pest.**
995. Major constituent of neem is **Azadiractin.**
996. Earthing up is used to manage of which pest in sugarcane **Early shoot borer.**
997. Which organism causes milky disease in Japanese beetle ***Bacillus papillae.***
998. Insecticide of plant origin **Nimbicidine.**
999. Sex pheromone is very effective against **Rice stem borer.**
1000. Dammer bee is ***Trigona iridipennis.***
1001. Spray fluid for ultra low volume spray/ha is **less than 5 lit.**
1002. An insecticide which belongs to Organophosphate group **DDVP.**
1003. Sting in worker bee is modified part of **Ovipositor.**
1004. *Micromus igorotus* is a **Predator.**
1005. *Tetranuera nigroabdominals* is a pest of **Finger millet.**

1006. Which one of the following is recommended for the control of mites **Dicofol.**

1007. Root grub adults emerges in the month of **May-June.**

1008. Freckled yellow of sorghum malady is transmitted by ***Perigrinus maidis*.**

1009. Bunchy top in sugar cane is due to the attack of **Top shoot borer.**

1010. The example for granular insecticide is **Carbofuron.**

1011. *Pentalonia nigronervosa* transmits the following disease **Katte of cardamom.**

1012. Diamond back moth is a pest on **Cabbage.**

1013. Citrus butter fly larvae infests **Leaves.**

1014. A pest which is found both in field and stored sweet potato is ***Cylas formicarius*.**

1015. *Microplitis maculipennis* is an effective parasitoid on **Castor semilooper.**

1016. The common name of *Oryzophilus surinamensis* is **Saw toothed beetle.**

1017. While feeding, thrust anterior part of body in to the fruit is the characteristic feature of ***Helicoverpa armigera*.**

1018. *Trichogramma* sp. parasitizes on the following stage of an insect **Egg.**

1019. Downward curling of leaves and petiole elongation in chilli is due to **Mites.**

1020. Sooty mould deposition on manage leaves is due to **Hoppers.**

1021. Coconut fronds showing characteristic V shaped clipping of leaves **Rhinoceros.**

1022. Bud necrosis disease is transmitted by ***Thrips palmi*.**

1023. *Colemenia spheneroides* is commonly called as **Deccan wingless grasshopper.**

1024. Bihar hairy caterpillar lays eggs in **Lower surface of broad leaved plants.**

1025. Control measures against the pest must be intiated at **ETL.**

1026. Red headed hairy caterpillar is an example for **Endemic pest.**

II. Indicate whether the following statements are True (T) or False (F)

- ___F___ Resistance to an insecticide is always dependent upon change in gene frequency compared to the Susceptible population.
- ___T___ Papaya ring spot virus is transmitted by an aphid and the best control strategy to manage the Problem is to develop the resistant varieties.
- ___T___ Insect mandibles can bite through metal and ovipositors of parasitoids can drill through hard wood because they are made up of zinc and manganese up to 25%.
- ___*F*___ In arrhenotoky only females are produced .
- ___*T*___ The legs of thrips end in vesicle claws.
- ___*F*___ In holopneustic type of tracheal system all the spiracles are closed and non functional.
- __*F*____ Food flow from mid gut to hind gut is regulated by stomodeal valve.
- ___*F*___ Silent spring revolutionized out thinking on pest management and resulted in the later development of the concept of IPM.
- ___*F*___ Cocoon, pupa and ootheca refer to the same developmental stage in insect.
- ___*T*___ Both adults and young ones of paurometabolous insects share the same habitat.
- ___*F*___ Insect flight is largely neurogenic thus allowing for extraordinary wing beat frequencies.
- ___*T*___ The brachypterous forms of BPH have low fecundity compared to macropterous forms.
- ___*F*___ The amount of insecticide to be applied varies with the type of sprayer.
- ___*F*___ Rodents show bait shyness to chronic poisons.
- ___*T*___ Insecticide synergists enhance toxicity by inhibiting detoxifying enzymes.
- ___*F*___ A pest resistant to DDT can show cross resistance to carbaryl.
- ___*T*___ In flowable formulations, use of organic solvents is minimized or avoided.

- ___F___ Chitinase is responsible for biosynthesis of chitin in epidermal cells.
- ___F___ Larvae of fruit sucking moth breeds on *Abutilon sp.*
- ___T___ *Ripiphorothrips cruentatus* makes corky layer on the grape fruits.
- ___F___ Growing of cow-pea as an intercrop to reduce the incidence of *Aprorema* modicella
- ___T___ The larvae of *Nymphula depunctalis* are adopted for semi-aquatic life.
- ___T___ Seed treatment of groundnut with chlorpyriphos 20 EC is recommended to manage white grub.
- ___T___ Red cotton bug lays eggs in the soil
- ___T___ *Scrobipalpa heliopa* causes unusual branching of seedlings in tobacco.
- ___F___ Insect flight is largely neurogenic thus allowing for extraordinary wing beat frequencies.
- ___F___ In terms of life-history strategies the *K*-selected species are the early colonizers and rapidly exploit patchyr esoruces.
- ___T___ Blood clotting in vertebrate hosts is reduced by the saliva of blood sucking insects that contains anti-thrombin, Factor X and vaso dialatory chemicals.
- ___F___ The haplo-diploidy hypothesis explains the evolution of social behaviour in termites, ants and bees.
- ___T___ One of the important postulates of Darwin's principle for evolution through natural selection is that organisms should produce more offspring than can survive.

III. Match the following from Column 'B' to Column 'A'

Column 'A'		Column 'B'
15	Pigeon pea sterility mosaic	1. *Heliciverpa*
19	Spider mites	2. Chitin synthesis inhibitor
13	False spider mites	3. *Achea janata*
9	Mango malformation	4. Antidote
17	Imidachloprid	5. JH activity
11	Methoprene	6. *Micromus* sp.
16	Flavonoides	7. AntiJH compound

6	Sugarcane wooly aphid	8. *Frankliniella schlutzei*
4	Atropine sulphate	9. *Aceria mangiferae*
14	Thiomethoxam	10. Phagocytosis
20	Chordotonal sensilla	11. JH Analogue
7	Precocenes	12. Pyrethroid
8	Groundnut stem necrosis	13. Tenuipalpidae
2	Buprofrezin	14. Actara
18	Cryoprotectant	15. *Aceria cajani*
12	Phenobarbitol	16. Allelochemicals
10	Plasmatocytes	17. Confedor
3	Bird perches	18. Glycerol
1	Bollgard	19. Tetranychiidae
5	Reproductive diapause	20. Vibration sensilla

OOO

CHAPTER 7

Plant Pathology

1. Phylum Plasmodiophoromycetes (Lower fungi) belong to kingdom **Protozoa.**
2. The cell wall of the mycelium of true fungi (Eumycota) contains **Glucan + chitin.**
3. *Mycosphaerella musicola* produces asci within **Locules.**
4. The teleomorphic stage of Cercospora is **Mycospaherella.**
5. Fungi belonging to order Sphaeropsidales produces conidia in **Pycnidium.**
6. Parasexuality was first discovered by **Pontecarvo.**
7. The total mass of nucleoprotein of different virus particles vary from **6 to 73 million Da.**
8. An example of virus containing ds DNA **Caulimovirus.**
9. The protein subunits of TMV consists of **158 amino acids.**
10. Multiplication of virus occurs in **Epidermal cells.**
11. In the peptidoglycon of the bacterial cell wall the two sugar molecules are connected by **Glycosidic bond.**
12. Porins are present in bacterial cell wall component **LPS layer**
13. The ribosomes size of bacteria is **70.**
14. A computer software system called Field Runner is used to **Sampling plants for disease.**
15. Isomerase enzyme patterns in fungal cell is used for **Differentiating strains (Biotypes).**
16. The computerized disease assessment training programme is **DISTRAIN.**

17. Measuring the size of the minute objects (spores) in a microscope is known as **Micrometry**.
18. The two flask techniques used in the extraction of nematodes is based on the principle of **Counter flow of water**.
19. Viruses enter the plant cells through the process of **Pinocytosis**.
20. Sugarcane ratoon stunting is caused by ***Cavibacter xyli* sub sp. *xyli***
21. Blotter test is used in **Seed health testing**.
22. Hyperplasia refers to **Increased cell division**.
23. An internally seed borne fungal disease **Loose smut of Wheat**.
24. An example of a soil fumigant **Methyl bromide**.
25. Significant contribution on the epidemiology of disease cycle of stem rust of Wheat was made by **K.C. Mehta**.
26. Moko disease of Banana is caused by ***Ralstonia solanacearum***.
27. Blast of Rice is caused by ***Magnaporthe grisea***.
28. The number of races of *Phytophothora infestans* prevalent in India is **82**.
29. Effective fungicide for the control of blast disease of Rice **Tricyclazole**.
30. Fungus with dichotomously branched sporangiophores at acute angles with curved pointed tips on which sporangia are borne belong to **Perenospora**.
31. Cleistothecium containing several asci bearing definite rigid appendages with coiled tips is characteristic of genus **Uncinula**.
32. Diseases which interferes with the conduction of water in plants **Wilts**.
33. Tumours/galls induced by *Agrobacterium tumefaciens* occur due to **Harmonal imbalance**.
34. Gemini viruses contain ss **DNA**.
35. Phytoplasmas contain **RNA+ DNA**.
36. Bacterial cell wall is made up of **Peptidoglycon**.
37. The genetic material in a bacterial cell is **A nucleoid region**.
38. Ribosomes in Prokaryotes **70 s**.
39. Agrocin-84 is a **Bacteriocin**.
40. International Mycological Institute is located at **England**.
41. Rots disease occur as a result of **Destruction and disintegration of affected tissue**.

42. Brine floatation method is followed to control **Tundu disease.**
43. MLO and Spiro plasma are mostly **Phloem inhibiting.**
44. Panama disease of banana is caused by ***Fusarium oxysporium* f sp. *Cubense*.**
45. Flagellation in Pseudomonas **Lophotrichous.**
46. *Hemelia vastatrix* causes **Coffee leaf rust.**
47. The number of races of *Hemelia vastatrix* recorded in the World **32.**
48. Internally seed borne fungal disease **Loose smut of Wheat.**
49. Internally seed borne bacterial disease **Black rot of cabbage.**
50. NEPO viruses are transmitted by ***Xiphinema index***.
51. Potato spindle tuber disease is caused by **Viroid**.
52. Most widely used fungicide for rust fungi **Plantvax.**
53. A biocontrol agent used to control root knot nematode ***Pasteuria***.
54. Streptomycin antibiotic affect **Protein synthesis.**
55. Hypersensitive reaction occurs due to **Accumulation of Phytoalexins.**
56. Toxin produced by *Fusarium* causing vascular wilts **Fusaricacid.**
57. The Indian Phytopathological society was started by **B.B. Mundkur**.
58. Purification of TMV in crystalline form by **Stanley**.
59. Monograph on the Genus Fusarium is authored by **C. Booth.**
60. Nematodes as plant parasites was reported for the first time by **Needham**.
61. Host plant on which Wheat rust fungus produces telial stage is known as **Primary host**.
62. NETU viruses are transmitted by ***Trichodorus***.
63. Stylet-borne viruses are **Non persistent**.
64. Who advanced the Gene for gene concept of disease and resistance and susceptibility **Flor.**
65. Fungi causing Downy mildew diseases are differentiated on the basis of **Sporangiophore character and germination of sporangia.**
66. Most of the Plant viruses contain **RNA**.
67. The famous Irish famine of 1842 was caused due to the outbreak of **Late blight of potato.**
68. Sulphur fungicides are very effective in the control of **Powdery mildews.**

69. Ratoon stunting disease of Sugarcane is caused by ***Clavibacter xyli*** **(Mollicutes).**
70. Ragi variety resistant to neck blast **GPU-28.**
71. Cysts of Golden nematode of potato can remain viable for **Up to 15 years.**
72. Bunchy top of Banana is transmitted by ***Pentalonia nigranervosa.***
73. A typical long cycled rust fungus produces during its life cycle **5 spore forms.**
74. *Puccinia graminis tritici* produces the following spore from on *Barberry vulgaris* **Pycnia + Aecia.**
75. The main source of inoculum for *Xanthomonas citri pv.citri* causing citrus canker **Infected twigs and old lesions on the standing tree.**
76. *Erwinia caratovora* sub. sp. *caratovora* causing soft rot of vegetables produces **Pectolytic enzymes.**
77. The number of races of *Ralstonia solanacearum* recorded in the World **5.**
78. *Agrobacterium tumefaciens* enters the plants through **Wounds.**
79. Systemic fungicide used for the control of Powdery mildew disease **Calixin.**
80. Late blight of potato can be controlled by spraying **Ridomil MZ.**
81. Seed treatment fungicide for the control Downy mildew disease of Sunflower **Apron 35 SD.**
82. Tomato leaf curl can be controlled by spraying **Triazophos (Hostathion).**
83. Blast of paddy is caused by ***Pyricularia oryzae.***
84. Biocontrol agent used for the control of Collar rot of plants caused by ***Rhizoctonia solani Trichoderma viridae.***
85. The term Virus was coined by **Beijerinck.**
86. Bacteriophage is **A Virus infecting bacteria.**
87. A virus particle is composed of **Nucleic acid+ protein.**
88. Bacteriophages are used for forecasting the incidence of **Bacterial blight of Paddy.**
89. Phytoplasmas are transmitted in nature through insect vectors **Leaf hoppers.**
90. Sulphur fungicides are very effective against **Powdery mildews.**

91. Solar heat treatment of seeds is recommended for the control of **loose smut of Wheat.**
92. Fungicide Metalaxyl is effective in controlling **Pythium and Phytophthora.**
93. Coffee leaf rust is caused by ***Hemelia vastatrix.***
94. Late blight of Potato is caused by ***Phytophthora infestans.***
95. *Ralstonia solanacearum* causes **Wilting.**
96. *Barbery vulgaris* is the alternative host for ***Puccinia graminis tritici.***
97. Burrowing nematode of Banana can be controlled by applying to soil **Furadon.**
98. Covering tomato nursery with nylon net is followed for the control of **Tomato leaf curl.**
99. A systemic fungicide used for the control of blast of Paddy **Beam.**
100. The primary source on inoculum for loose smut of Wheat **Internally seed borne hypha.**
101. Primary source of inoculum for Angular leaf spot of cotton **Seed borne bacterium.**
102. *Agrobacterium tumefaciens* induces **Galls/tumours.**
103. Effective control for Katte disease of Cardamom is **Eradication and burning of affected clumps.**
104. Khaira disease of Paddy is due to the deficiency of **Zinc deficiency.**
105. *Pasteuria penetrans* is used as a biocontrol agent for **Root knot of tomato.**
106. The sexual spore produced by Downy mildew fungus is **Oospore.**
107. Antibiotic used to control bacterial diseases of plants **Streptocycline.**
108. Cell wall of fungi belonging to ascomycetes contains **Chitin.**
109. The number of ascospore produced in a typical asci are **8.**
110. Black mold fungus usually refers to ***Aspergillus niger.***
111. Fungal nucleolus consist mostly of **RNA.**
112. The perfect stage of *Aspergillus nidulans* is **Emerincella.**
113. Demi cyclic rust fungus lack **Uredial stage.**
114. From class Deuteromycetes contain the following number of subclasses **3.**
115. Fungus producing conidia either in Pycnidia or acervuli belong to **Coelomycetide.**

116. Dictyospora is a spore with **Both horizontal and vertical septa.**
117. The first expert (Advisory) system in Plant Pathology was developed in USA in the year **1983.**
118. The maximum tolerance level for potato leaf roll virus in certified seed potatoes is **0.5-1.0%.**
119. A fungicide which does not belong to Triazole group is **Ethazol.**
120. Powdery mildew diseases are effectively controlled by spraying **Rubigon.**
121. K 84 is a **Bacteriocin.**
122. A sterol inhibiting fungicide is **Prochloroz.**
123. Benzimidazole fungicides interfere with **Nuclear division.**
124. The teliospores of *Sphacelotheca sorghi* causing grain smut of sorghum can remain viable for **13 years.**
125. Fungi with coenocytic hyaline mycelium producing sporangiophores. similar to vegetative hyphae or if different than of indefinite growth belong to family **Pythiaceae.**
126. The total number of amino acids in the protein subunits of TMV consists of **158 amino acids.**
127. The total mass of nucleoprotein of different virus particles vary from **4.6-73 million.**
128. MLO's (Phytoplasmas) in plant system were translocated through **Phloem sieve tubes.**
129. Caulimovirus contain **ds DNA.**
130. The total amount of DNA in the chromosomes of a bacterium *viz Escherichia coli* is about **4700 kilobase.**
131. The thickness of the plasma membrane of bacterial cell is **8 nm.**
132. N-Acetyglucosaminoacid and N-acetyl muramic acid in the peptidoglycon unit is linked by **1, 4- glycosidic bond.**
133. Approximate threshold density of *Globodera rostochinensis* for causing damage to potato crop **31 cysts/100 g of soil.**
134. An example of a migratory endoparasite is **Radopholus.**
135. Genus Alternaria belong to family **Dematiaceae.**
136. The phenomena of parasexuality was first discovered by **Pontecarvo.**
137. The protein content in a bacterial cell on dry weight basis is **60%.**
138. The diameter of transmembrane protein in a bacterial cell is **1 nm.**

139. Bergeys manual of Determinative Bacteriology is published in **4 volumes.**
140. The protein associated with the genetic transformation process in bacteria is **Rec A protein.**
141. Plant pathogenic bacteria which does not belong to coryneform group is **Pseudomonas.**
142. LOPAT test is used for the identification of **Fluorescent Pseudomonas.**
143. Ti Plasmid is present in ***Agrobacterium tumefaciens***.
144. Turnip shaped single celled pedicellate teliospore borne singly is a characteristic of genus **Hemilia.**
145. Bakane disease of rice caused due to the production of any one of the following by the fungal pathogen **Gibberellins.**
146. Plant disease which affect the absorption and accumulation of water and minerals **Wilts.**
147. The incubation period for stem rust of wheat in South India is **one month.**
148. An Antibiotic that affecting nucleic acid synthesis is **Griseofulvin.**
149. Oxanthiin fungicides act on the fungal cell by **Inhibiting mitochondrial respiration.**
150. The immersion oil used in the compound microscope has a refractive index of **1.515.**
151. Most viruses move longer distances through the plants in the **Plasmodesmata.**
152. Virus with split genome **Wound tumour virus.**
153. The number of days required for most of the viruses to move out of inoculated leaf is **2 to 5 days.**
154. The rice tungro virus contains **ds DNA.**
155. The nematode organ which have important sensory function is **Amphids.**
156. Nematodes with two ovaries are known as **Didelphic.**
157. Bacteria as causal agent of plant disease was first reported by **T.J. Burill.**
158. Fungi causing smut diseases belong to **Basidiomycotina.**
159. The number of spore types produced by *Puccinia graminis tritici* during its life cycle is **5.**

160. Brown rot of Potato is caused by ***Ralstonia solanacearum***.
161. Hinosan is used to control **Blast of Paddy**.
162. NEPO viruses are transmitted by nematode genera **Xiphinema.**
163. Alternate host for *Puccinia graminis tritici* ***Barberry vulgaris***.
164. Angular leaf spot of cotton is caused by ***Xanthomonas campestris pv.malvacearum***.
165. A disease occurring severely and confined to small locality /area is **Endemic disease.**
166. A fungal antagonist used in the biological control of soil borne fungal diseases ***Trichoderma*** sp.
167. Effective fungicide for the control of late blight of potato **Ridomil MZ.**
168. *Colletotrichum falcatum* causes **Red rot of sugarcane.**
169. Koleroga of arecanut is controlled by spraying **Bordeaux mixture.**
170. Tomato leaf curl virus is transmitted by ***Bemisia tabaci***.
171. Leaf shredding and green ear symptoms are noticed in **Downy mildew.**
172. The genetic material in a typical bacterial cell is contained in **Chromosomal strand.**
173. Katte disease of cardamom is transmitted by ***Pentalonia nigranervosa***.
174. Citrus greening disease is caused by a **Bacteria like organism (Spiroplasm).**
175. A systemic fungicide used for the control of Powdery mildew diseases **Tridemorph (Calixin).**
176. *Rhadopholus similis* infecting Banana can be controlled by **Paring and prolinage treatment.**
177. A systemic fungicide used for the control of rust diseases **Plantvax.**
178. Black rot of cabbage is caused by ***Xanthomonas campestris pv. campestris***.
179. Tikka disease of Groundnut can be controlled by spraying **Bavistin.**
180. Swollen shoot of cocoa is transmitted by ***Pseudococcus njalensis***.
181. Burrowing nematode of banana is **Migratory**.
182. Stem rust of wheat is caused by *Puccinia graminis tritici* has been successfully cultured on medium by **Williams.**

183. Cell wall of fungi belonging to Oomycetes contain **Cellulose.**
184. White tip of rice caused by *Aphelenchoides besseyi* and is reported by **Dastur.**
185. *Amomum subulatum* is one of the alternate host for **Katte disease of cardamom.**
186. Certain bacteria to gain entry into host tissues to dissolve cell wall of the plant it produce **Pectolytic enzyme.**
187. Nuclear position of secondary mycelium of basidiomycotina is **n+n.**
188. *Rhizoctonia solani* a root rot causing pathogen comes under order **Mycelia sterilia.**
189. In case of *Uncinula nicator* ascocarp consists of more number of asci appendages type is **Coiled type.**
190. In 1927 discovered the functions of Spermatia and their role in dikaryotization by **Craigie.**
191. One of the important cultural method suggested for eradication of *Fusarium oxysporum fsp. cubense* from soil is **Flooding the soil.**
192. Anthracnose of beans caused by ***Colletotrichum lindemuthianum***.
193. *Viscum* sp. is a plant parasite **Partial parasite**.
194. Insect vector transmitted viruses are more serious during **Summer season.**
195. Tip over disease of banana is caused by ***Erwinia caratovora***.
196. A major threat for pomegranate cultivation in recent days is bacterial blight and it is caused by ***Xanthomonas axonopodis pv. punicae***.
197. 1882-85 discovered Bordeaux mixture by **P.M.A. Millardet.**
198. Citrus tristeza could be effectively managed by **Cross protection.**
199. Metalaxyl is a specific fungicide against **Perenosporales.**
200. Plant pathogenic bacteria are **Heterotrophs.**
201. *Puccinia graminis tritici* produces aeciospores on **Barberry plants.**
202. Turnip shaped single celled pedicellate teliospore borne singly is characteristic of **Hemelia.**
203. Viroid- a new infectious agent causing potato spindle tuber disease was first reported by **Diener.**
204. Sexual spore produced by fungi causing downy mildew diseases **Oospore.**

205. Wilt disease of banana caused by *Ralstonia solanacearum* is called as **Moko disease of banana.**

206. Bursting of Pseudostem in banana is observed in plants suffering from **Panama wilt.**

207. The primary source of inoculum for loose smut of wheat **Internally seed borne mycelium.**

208. Nylon netting of Tomato nursery is practiced for the management of **Leaf curl.**

209. Late blight of potato is effectively controlled by **Metalaxyl 0.1%.**

210. A computer software system developed for field sampling of diseased specimens **Field Runner.**

211. A computerized disease forecasting system for the occurrence of late blight of potato **Disease.Pro.**

212. Pathogens causing soft rot diseases chiefly produce **Pectolytic enzymes.**

213. Blotter test is commonly followed for the detection of **Seed borne fungi.**

214. The alternate host for *Hemilia vastatrix* is **Not known.**

215. Most widely used bio control agent is ***Pseudomonas fluorescence.***

216. Groundnut bud necrosis virus is transmitted through **Thrips.**

217. Most widely used fungicide for smut fungi **Vitavax.**

218. The sexual spores produced by fungi belonging to Deuteromycotina are **No sexual spores are produced.**

219. MLO's are sensitive to **Tetracycline.**

220. An example of bioagent is ***Trichoderma viridae.***

221. Tundu disease of Wheat is caused by ***Anguina tritici*** and ***Corynebacterium tritici.***

222. Outbreak of a disease over a large plant population is known as **Epidemic.**

223. Black stem rust of wheat is caused by ***Puccinia graminis tritici.***

224. Peach leaf curl is caused by **Fungus.**

225. Non-motile spores of fungus are called **Aplanospores.**

226. Teliospore is a **Sexual.**

227. Grain smut of Sorghum is caused by ***Sphacelotheca sorghi.***

228. TMV is transmitted through **Seed.**

229. Coconut cadang-cadang disease is caused by **Viroid.**
230. In stinking smut of wheat, a fishy odour is due to **Trimethylamine.**
231. The most important soil borne disease in nursery is **Damping off.**
232. Tomato leaf curl disease is transmitted by **Whitefly.**
233. Matured virus particle is called **Virion.**
234. A rust fungus that completes its entire lifecycle on single host is called as **Autoecious.**
235. Mutualistic association of fungus with plant parts is **Mycorrhiza.**
236. An example for double stranded DNA virus is **Cauliflower mosaic virus.**
237. The loss of turgidity and drooping of leaves or shoots caused by fungi and bacteria **Wilt.**
238. *Puccinia graminis tritici* infects **Barbery and Wheat.**
239. Chemical that halts the progress of a fungus is known as **Fungistat.**
240. Perithecium is a sexual fruiting body produced by ***Erysiphe polygoni.***
241. Cleistothecium is a **Closed fruiting body.**
242. Rhodamine B is used to detect viruses in **Fluorescent antibody technique.**
243. Late blight of potato is the best example for **Pandemic disease.**
244. The sequence of fixing components in the well in ELISA are **Antigen-antibody –conjugate.**
245. Which of the facts related to *Hemilia vastatrix* is incorrect? ***Coffea robusta* is partially resistant.**
246. Which fact with refrence to nematodes is not true? **Nematodes enter plants by direct penetration only.**
247. Which fact related to Tabtoxin is not true? **Tabtoxin is produced by *Alternaria helianthi*.**
248. The toxin that is host specific is **Victorin.**
249. The induced structural defences is **Cytoplasmic defence reaction.**
250. Which fact with reference to vertical resistance is not true? **Varieties with complete vertical resistance remain resistance forever.**
251. Which forecasting is best for the late blight of potato? **Forecasting system based on weather conditions.**

252. The mycologist involved with late blight of potato is **Anton de Bary.**
253. The fungus involved with foot rot of pepper is ***Phytophthora capsici.***
254. The genus of powdery mildew pathogen that produces both external and internal mycelium is **Phyllactinia.**
255. The fungal genus that causes anthracnose diseases in crops is **Glomerella, Elsinoe and Colletotrichum.**
256. Powdery mildew disease predominantly occurs during **Winter.**
257. The primary source of infection for black rot of cabbage is **Seed.**
258. The sexual fruiting bodies produced by fungi are **Perithecium.**
259. The quarantine act passed by India is **Destructive insect and pest act 1914.**
260. Citrus greening is transmitted by ***Toxoptera citricidus.***
261. Father of modern Plant Pathology **Anton de Bary.**
262. Germ theory was originated from the works of **Louis Pasteur.**
263. The first person to transmit virus infection from infected to healthy plant is **Mayer.**
264. The autoclaving temperature at 15 lb psi is **121.60°C.**
265. The chemical reaction involved in Bordeaux mixture is **$Cuso_4 + Ca(OH)_2 = Cu(OH)_2 + CaSo_4$.**
266. Bordeaux mixture 0.5% is prepared by mixing copper sulphate and lime water in the proportion **0.25:0.25: 50.**
267. The conidial shape of *Pyricularia oryzae* is **Pyriform.**
268. Green ear is the characteristic symptom produced by ***Sclerospora graminicola.***
269. The sclerotial bodies of *Claviceps microcephala* upon germination produce the following spores and fruiting bodies **Ascospores in perithecia.**
270. Seed treatment with a fungicide is not useful against **Head smut of.**
271. Which fact with reference to *Cercospora arachidicola* is not true? **Conidia shorter than *Cercospora personata*.**
272. Which statement with reference to *Phytophthora* sp. is incorrect? ***P. capsici* causes fruit rot in areca.**
273. *Elsinoe ampelina* causes anthracnose in **Grapes.**
274. Pick a copper based fungicide **Phytolan.**

275. Which is the most effective chemical to control downy mildew in maize? **Ridomil MZ.**
276. Point out the non systemic fungicide **Ziram.**
277. Which is the most effective soil drenching fungicide? **Thiram.**
278. The trade name of the fungicide Triademefon is **Bayleton.**
279. The trade name for Oxycarboxin is **Plantvax.**
280. The mycelium of Oomycetes fungi is **Coenocytic.**
281. Repeating spores of rust fungi is **Urediospore.**
282. Secondary mycelium of Basidiomycotina is **Dikaryotic.**
283. *Macrophomina phaseolina* belongs to the order **Sphaeropsidales.**
284. Late blight of potato caused famine in Ireland in year **1845.**
285. Frog eye spot of tobacco is a serious disease caused by ***C. nicotianae*.**
286. The perfect stage of *Helminthosporium oryzae* is ***D. oryzae*.**
287. Citrus canker caused by *Xanthomonas citri* has been suggested to be originated from **Java and India.**
288. Viroids are devoid of **Protein coat.**
289. Leaf curl of tomato is a geminivirus whose nucleic acid is made up of **DNA.**
290. Contagium vivum fluidum theory was proposed by **Beijerinck.**
291. Papaya ring spot virus is a non persistent virus transmitted by ***Myzus persicae*.**
292. Of the 25 virus disease recorded in India on Potato, the most destructive one is **Rugose mosaic.**
293. Yellow mosaic of Bhendi and mosaic disease of Cassava are transmitted by ***Bemisia tabaci*.**
294. Trade name of Chlorothalonil is **Daconil.**
295. Curzate, a systemic fungicide is specific to **Downy mildew.**
296. The primary source of inoculum for stem and fruit canker of tomato is **Seed borne only.**
297. The genome of Tomato leaf curl virus is **ss DNA.**
298. A typical symptom of bird's eye spot is caused by ***Clavibacter michiganense sub*** sp. ***michiganense*.**
299. Sandal spike was first reported from India in the year **1898.**
300. Little leaf of Brinjal is caused by **Phytoplasma.**
301. A fungus causing wart of potato is **An obligate.**

302. A long flexious filamentous ss RNA virus is **Beet yellow virus.**
303. The fruiting body of *Colletotrichum capsici* is **Acervulus.**
304. A toxic substance produced by *Trichoderma viridae* is **Viridin.**
305. The famous Irish famine occurred in the year **1845.**
306. The resistant structure produced by *Fusarium* sp. is **Chlamydospores.**
307. Tomato spotted wilt virus is transmitted by **Thrips.**
308. Bacterial canker of tomato is caused by ***X. campestris. pv. malvacearum*.**
309. Phytoplasmas are sensitive to antibiotics of **Tetracycline.**
310. The type of nucleic acid present in yellow mosaic of Bhendi is **ss DNA.**
311. Powdery mildew of chilli is caused by ***Leveillula taurica*.**
312. Pea rust is caused by **Uromyces.**
313. Common scab of potato is caused by **Fungi.**
314. Black spot of Rose is caused by ***Diplocarpon* sp.**
315. A disease caused by *Colletotrichum circinans* in onion is **Smudge.**
316. Fungi belonging to the order Ustilaginales are known to cause **Smuts.**
317. A button like structure which penetrate the host and draws nutrients **Haustoria.**
318. A plant viral disease of seed borne nature **Common bean mosaic virus.**
319. A disease affecting food conduction in plant **Vascular wilt.**
320. A nematode which is of migratory endoparasite **Radopholus.**
321. Who was considered as founder of virology? **Beijerinck.**
322. Sporangia are produced in chains **Albugo.**
323. Ostiolate ascocarp is known as **Perithecium.**
324. A bacterial genus which enter through nectarines is **Erwinia.**
325. An endophytic powdery mildew genus having the cleistothecial characters like indefinite myceloid appendages **Leveillula.**
326. A tuft of flagella one or both ends is called **Amphitrichous.**
327. Prokaryote without the cell wall **Bacteria like organisms.**
328. Low molecular weight infectious RNA without a protein envelope is **Viroid.**
329. Robigalia was celebrated by Romans against **Rusts.**

330. Khaira disease of paddy is caused by **Zn.**
331. A partial stem parasite of trees **Cuscuta.**
332. Cross protection can be achieved by inoculating the plants first with **Mild strain of same virus.**
333. Use of antibiotics in plant disease control is by **M.J. Thirumalachari.**
334. Highly contagious virus **Tobacco mosaic virus.**
335. Dikaryotic mycelium is a distinct feature of fungi belonging to **Basidiomycotina.**
336. Bio-agent used in the management of airborne diseases ***Trichoderma viridae*.**
337. A disease incited by an endosedentary plant parasitic nematode **Meloidogyne.**
338. Plant pathogenic bacteria having relevance in genetic engineering **Agrobacterium.**
339. PCR stands for **Polymerase chain reaction.**
340. Xanthomonas penetrates the host tissue through **Stomata.**
341. Practical and feasible method in managing soil borne diseases is through **Use of disease resistant varieties.**
342. A systemic fungicide is **Carbendazim.**
343. An infectious stage of plant parasitic nematode is **II stage.**
344. All plant pathogenic bacteria are **Facultative parasites.**
345. A culturable cell wall less fastidious prokaryote causing plant diseases is **Phytoplasma.**
346. Mycelium of class Oomycetes is always **Coenocytic.**
347. Streptomycin sulphate, an antibiotic was isolated by **Waksmann.**
348. Tur wilt is caused by ***Fusarium udum*.**
349. Race specific resistance is called **Vertical resistance.**
350. L.C. Coleman was a general pathologist during 1914 but worked and suggested control measures for **Koleroga of Areca.**
351. *Exobasidium vexans* which causes **Blister blight of tea.**
352. Early blight of potato and tomato caused by ***Alternaria*.**
353. Downy mildew causing pathogens, powdery mildew causing pathogens and rust causing pathogens are **Obligate.**
354. The phanerogamic parasite with the Sorghum crop **Striga.**

355. Teliospores are single celled, papillate and stalked, the genus identified as **Uromyces.**
356. One of the chemical weapons secreted by the pathogen in production of disease in plants **Phytotoxins.**
357. Puccinia is a member of the order **Uredinales.**
358. Conidia are produced singly at tips of conidiophores by **Leveillula.**
359. When plant showed the partial resistance against all the races of pathogen then it is a type of **Horizontal resistance.**
360. The teliospore of coffee leaf rust pathogen is **Turnip shaped.**
361. What cannot be detected by ELISA technique? **Viroid.**
362. Viruses are **Nucleoproteins.**
363. Monogenic resistance is also referred to as **Vertical resistance.**
364. In a long cycled rust "0" stage refers to **Pycnium.**
365. Cross protection is effective in controlling **Citrus tristeza.**
366. Pathogen inducing soft rot symptoms primarily produce **Pectolytic enzymes.**
367. Dicytospora is a spore with **both vertical and horizontal septa.**
368. Genus Alternaria belongs to family **Dematiaceae.**
369. The phenomena of parasexuality was first discovered by **Pontecarvo.**
370. The protein content in a bacterial cell on dry weight basis is **60%.**
371. Virus with split genomes **Wound tumour virus.**
372. The most widely used granulated formulations have granule size of **0.2 to 1 mm.**
373. The additive chemicals which possess surface modifying properties are called as **Surfactants.**
374. Components of Chestnut compound are **Ammonium carbonate.**
375. The great Bengal famine in 1943 was due to **Brown leaf spot of Rice.**
376. Phytoalexin inducers viz ferric chloride, nickel nitrate etc., offer protection against **Blast.**
377. False smut of Rice is caused by ***Ustilaginoidea virens.***
378. The causal agent of foot rot or Bakane disease of rice produces **Gibberellic acid.**
379. Red rot of Sugarcane is caused by ***Colletotrichum falcatum.***
380. The causal agent of Grassy shoot disease in sugarcane is caused by **Phytoplasma.**

381. Stunting disease of sugarcane is caused by ***Clavibacter xyli subsp. xyli***.
382. Soil drenching with 0.4% bleaching powder is effective in **Leaf scald**.
383. Bacterial blight or black arm of cotton is incited by ***Xanthomonas campestris pv.malvacearum***.
384. Sudden and complete wilting in cotton is due to **both Root rot by *Rhizoctonia solani* and root rot by *Rhizoctonia bataticola***.
385. Apple scab is caused by ***Venturia inaequalis***.
386. Citrus exocartis is transmitted through **Both knife and bud wood**.
387. Downy mildew of grapes caused by ***Plasmopora viticola***.
388. Double stranded DNA isometric virus is **Cauliflower mosaic virus**.
389. Plant viruses move from cell to cell through **Plasmodesmata**.
390. Viruses are **Nucleoprotein**.
391. Mollicutes are **Sensitive to Tetracycline**.
392. In 1845, the late blight of potato destroyed the potato crop of Ireland was caused by ***Phytophthora infestans***.
393. In 1943, Bengal had faced a serious famine which cause a great loss in rice yield caused by ***Helminthosporium oryzae***.
394. Who is the father of Plant Pathology? **Anton de Bary**.
395. Who described the first plant nematode disease, the seed gall caused by *Anguina tritici* in 1743 A.D? **Needham**.
396. For the first time who reported that a plant disease (Fire blight of pear) was caused by bacterium (*Erwinia amylovora*) in 1882 **T. J. Burill**.
397. The first Indian scientist who collected and identified fungi in India **K. R. Kirtikar**.
398. 'Fungi and disease in Plants' was written in 1918 by **E. J. Butler**.
399. 'Fungi and Plant Diseases' was written by **B. B. Mundkur**.
400. Select the wrong pair **Atrophy: inhibition of growth of dwarfing**.
401. A bacteria which have a large number of flagella all over the cell is categorized in **Peritrichous**.
402. In bacteria, variability is caused by **Conjugation, Transformation and Transduction**.
403. The scientific name of potato black leg pathogen is ***Erwinia carotovora***.

404. J. C. Luthra and his associates developed the solar heat treatment of wheat seeds for the control of **Loose smut**.

405. Citrus canker (Lesions) which originated from China is caused by pathogen ***Xanthomonas campestris pv. citri***.

406. Select the wrong pair **Ergot of Bajra: *Erwinia carotovora***.

407. Common scab of potato can be controlled by reducing pH below 5.2 is caused by the pathogen ***Streptomyces scabies***.

408. Tundu disease (Yellow ear rot) of Wheat caused by *Clavibacter tritici* is usually associated with ***Anguina tritici***.

409. Sandal spike disease of sandal (*Santalum album*) is caused by **MLO**.

410. The adjacent cells of the same fungal hypha brought into communication with each other by means of loops is known as **Clamp connection**.

411. What are the sexual spores of fungi **Zygospores**.

412. How many ascospores are enclosed in the mother cell, the ascus in ascomycetes **8**.

413. The perfect (Sexual) stage not seen in **Deuteromycotina**.

414. The black wart disease of potato specially confined to Darjeeling hills is caused by the pathogen ***Synchytrium endobioticum***.

415. 'White blisters of crucifers is caused by the pathogen ***Albugo candida***.

416. In uredinales basidia usually bearing how many numbers of basidiospores **4**.

417. The spores which are produced typically by the binucleate mycelium and mainly function as repeating spores for the spread of rust disease in wheat is called **Urediospores**.

418. Promycelium in rust fungi bears **Basidiospores**.

419. The disease damping off seedlings during nursery stage is caused by the pathogen ***Pythium aphanidermatum and Pythium debaryanum***.

420. The bacterial colony is known as **Ooze**.

421. The gummosis disease of citrus which is caused by *Phytophthora palmivora* is identified by observing **Oozing of gum**.

422. *Albugo candida* causes white blisters or white rust of crucifers is a an **Obligate parasite**.

423. *Albugo candida* produces **Basidiospores.**
424. The powdery mildew of wheat is associated with fungi ***Erysiphe graminis*** sp. ***tritici***.
425. The disease caused by *Leptosphaeria sacchari* in sugarcane is **Ring spot.**
426. Select the disease in which small droplets of pinks or light honey coloured fluid (The honey dew stage) exudes from the spikelets **Ergot of bajra.**
427. Which fungicide gives effective control of Covered smut of barley? **Vitavax.**
428. The plants produce a whip like black shoot, often very long and much curved on itself in smut of sugarcane is caused by pathogen ***Ustilago scitaminea***.
429. The false smut of sugarcane can be controlled by **Avoid the practice of rationing, Disinfection of setts before planting and Removal of smutted whips from the field.**
430. The Karnal bunt of wheat is caused by ***Neovossia indica***.
431. The causal organism *Neovossia indica* produces **Teliospores.**
432. The causal organism of bunt of rice is ***Tilletia foetida***.
433. The effective control of flag smut of wheat (*Urocystis tritici*) can be done by adopting of **Seed treatment.**
434. Which type of spores of *Puccinia graminis tritici* infect the barberry plant **Basidiospores.**
435. In rust cycle the cereal host is infected by **Teliospores.**
436. The yellow rust of wheat is caused by ***Puccinia striiformis***.
437. Rust of linseed and flax (autoecious rust) is caused by ***Melampsora lini***.
438. Early blight of potato produces **Conidia.**
439. Early blight of potato is **Soil borne disease.**
440. Leaf spot or tikka disease of groundnut is/are caused by ***Cercospora arachidicola and Cercospora personatum***.
441. Brown leaf spot disease of rice is caused by ***Drechslera***.
442. The pathogen of rice blast or rotten neck is ***Pyricularia oryzae***.
443. Tea rust is caused by **Algae.**
444. The *Colletotrichum falcatum* produces **Conidiospores.**
445. Wilt of pigeon pea is caused by ***Fusarium udum***.

446. The most important symptom of wilt of cotton (caused by *Fusarium oxysporum* sp. *vasinfectum*) is **Discolouration of tissues and plugging of vessels by hyphae.**
447. The pathogen responsible for charcoal rot of soybean is ***Macrophomina phaseolina*.**
448. The scientific name of burrowing nematode is ***Radopholus* sp.**
449. The nematodes are **Triploblastic.**
450. The nematodes lack organs for **Circulation and Respiration.**
451. In the attack of which class of nematodes, presence of white. to brown cysts projecting on the root surface is a characteristic symptom ***Globodera*.**
452. Select the nematode(s) which act as vectors of plant viruses ***Xiphinema index*, *Longidorus* and *Trichodorus*.**
453. Citrus greening caused by the **Fastidious bacteria.**
454. The uredospores of *Puccinia graminis* are disseminated by **Wind.**
455. Bacterial cell division mainly by **Binary division.**
456. Cell wall lack microorganism **MLO.**
457. MLO first discovered by **Doi.**
458. Plant Pathology written by **R. S. Singh.**
459. In north India and central India, the black rust inoculum cause from **Hilly area.**
460. Rice blast pathogen perfect stage is ***Magnaporthe grisea*.**
461. Smut of maize is caused by ***Ustilago maydis*.**
462. Virus capsid is made up of **Protein.**
463. Phyllody disease of sesamum spread by **Leaf hopper.**
464. Fungi imperfecti includes in **Deuteromycotina.**
465. Rust includes in **Basidiomycotina.**
466. Smut includes **Basidiomycotina**
467. Powdery mildew include in **Ascomycotina**
468. Downy mildew include in **Oomyctes**
469. Downy mildew of pea caused by ***Perenospora pisi***
470. Father of Indian Mycology **E. J. Butler.**
471. Renowned scientist Norman E. Borlaug belongs to **Plant Pathology.**
472. Which of the following bacteria generally used in genetic engineering for transfer of character from one organism to another ***Agrobacterium tumefaciens*.**

473. Black heart is a physiological disorder of **Potato.**
474. Which of the following is not correctly matched? **Wheat-Red rust.**
475. Tungro disease of rice is spread by ***Nephotettix virescense*.**
476. Panama wilt is a disease of **Banana.**
477. VAM is **Fungi.**
478. The major storage fungi that affects the food grain is **Aspergillus.**
479. Yellow leaf mosaic of bhendi crop spreads by **Whiteflies.**
480. Heterodera avenae is **Cyst nematode.**
481. Which one of the following microbial agents is being commercially exploited as biocontrol agent? ***Trichoderma viride*.**
482. Which fungicides is not systemic in nature **Thiram.**
483. Entomopathogenic fungi are ***Beauveria bassiana* and *Metarrhizium anisopliae*.**
484. Little leaf in brinjal is caused by a **Mycoplasma.**
485. Bacterial diseases are controlled by use of chemicals **Antibiotics.**
486. Application of potash increases **Disease resistance in plants.**
487. The root-knot nematode generally have interaction with ***Ralstonia solanacearum*.**
488. Margosa is a highly effective product in reducing root-knot nematode population belongs to ***Azadirachta indica*.**
489. Viruses contain **either RNA or DNA, never both.**
490. The viruses which are usually helped or accompanied by smaller spherical particles of another serologically unrelated virus is known as **Satellite virus.**
491. The fungi which transmit plant viruses belong to class **Plasmodiophoromycetes.**
492. For quick and accurate detection of viruses can be done by **ELISA.**
493. Most of the plant viruses transmitted by **Aphids.**
494. Bunchy top of papaya (caused by MLO) is spread by **Leaf hoppers.**
495. Potato spindle tuber disease is transmitted by **Mechanically.**
496. Suicidal germination takes place in **Striga.**
497. Dinocap (methyl heptyl dinitrophenyl crotonate) is sold in market as **Karathane.**
498. Which systemic fungicide was first time discovered by Von Schleming and Kulka? **Oxathiin.**

499. Select the nematicides group which are non-fumigant in nature **Nemagon (DBCP).**
500. Nematicides inhibits which enzyme of nervous system in killing of nematodes **Acetyl cholinesterase.**
501. Major genes contributes disease resistance in the form of **Vertical resistance.**
502. A virus that infects bacteria and destroys them is called as **Bacteriophage.**
503. The science which deals with cause of the plant diseases and nature of the causal agent **Etiology.**
504. The sexual, thick walled and resting spores of the rust fungi is known as **Teliospores.**
505. Who discovered the downy mildew for the first time in India? **K.R. Kirtikar.**
506. Which one of the following is a single cell fungi? **Yeast.**
507. What is the source of Agar Agar? **Algae.**
508. Fungi belonging to order Meloconiales produce conidia within **Acervulus.**
509. The first Phytoalexin Pisatin was isolated and purified by **Cruickshank and Perrin.**
510. White rust of crucifers is caused by ***Albugo candida*.**
511. Captan as fungicide was introduced by **Kittleson.**
512. 1% Bordeaux mixture is a combination of **1kg copper sulphate + 1kg Lime + 100 lt water.**
513. Bordeaux mixture was an accidental discovery for **Downy mildew of grapes.**
514. Metalaxyl is specific against **Oomycetes.**
515. Domestic quarantine against **Golden nematode and Wart of potato.**
516. Oxanthiin group of systemic fungicides are effective against **Smuts and Rusts.**
517. Organic sulphur fungicides are derived from **Dithiocarbamic acid.**
518. A typical nematode life cycle consists of **6 stages.**
519. A plant pathogenic bacteria peritrichously flagellate **Erwinia.**
520. Gram +ve plant pathogen ***Streptomyces*.**

521. Variability in bacteria is brought about by **Conjugation, transformation, transduction.**

522. Crown gall of apple is due to ***Agrobacterium rhizogenes***.

523. Common blight of beans is due to ***Xanthomonas campestris pv. phaseoli***.

524. Bacterial soft rot of vegetables is due to ***Erwinia caratovora pv. caratovora***.

525. Physical properties of virus are **Dilution end point, thermal inactivation point, longevity invitro.**

526. Fungus transmitting tobacco necrosis is ***Olpidium brassicae***.

527. Nematode transmitting grape fan leaf virus ***Xiphinema index***.

528. Tomato spotted wilt virus is transmitted by ***Thrips tabaci***.

529. Potato leaf roll virus is transmitted by ***Myzus persicae***.

530. Veinal clearing and veinal chlorosis of leaf is the chief symptom of the disease **Bhendi yellow vein mosaic.**

531. Citrus variegated is due to **Virus.**

532. Existence of formae specialis was first demonstrated by **Erikson.**

533. Main source of uredospores of *Puccinia graminis tritici* are **Foot hills of Himalayas.**

534. Primary inoculum of brown and yellow rust comes from **Foot hills of Himalayas.**

535. Rice Tungro virus has two kinds of particles which are **Bacilliform and spherical/isometric.**

536. Rice Tungro virus is transmitted by **Leaf hoppers.**

537. The cause of Rice Yellows disease is a **Phytoplasma.**

538. The causal organism of Bacterial leaf blight of Rice is ***Xanthomonas campestris pv. oryzae***.

539. The first plant parasitic nematode reported was ***Anguina tritici* by Needham.**

540. Journal of Nematology published in India **Indian journal of Nematology.**

541. *Pasteuria penetrans* is used as a biocontrol agent for managing **Meloidogyne.**

542. Paring and prolinage are done to manage **Banana decline.**

543. Coconut cadang cadang is due to **Viroids.**

544. Citrus greening is due to **Phloem limited fastidious vascular bacteria.**
545. The founder of mycology was **P.A. Micheli.**
546. Species Plantarum is the greatest contribution of **Linnaeus.**
547. With regard to nutrition Erysiphe is **Ecto Parasite.**
548. The class ends with **Mycetes.**
549. Pythium belongs to class **Oomyctes.**
550. Pycnidia are formed in order **Sphaeropsidales.**
551. Anterior tinsel flagellum is characteristics of **Chytridiomycetes and Hyphochytridiomycetes.**
552. B.B. Mundkur wrote **Fungi and Plant diseases.**
553. Which one of the following is known as Downy mildew fungus **Sclerospora.**
554. Rust teleutospore on germination produce **Four basidiospores.**
555. Fungi and Disease of plant was written by **B.B. Mundkur.**
556. Used lens for the first time to study microorganism **Antony Van Leeuwenhoek.**
557. Domestic Quarantine exist on **Wart of Potato.**
558. Once the pathogen has established in an area, steps taken to remove or destroy is known as **Eradication.**
559. Loose smut of wheat can be controlled by **Seed treatment.**
560. Flag smut of wheat is caused by ***Urocystis tritici.***
561. Agallol is used for **Seed treatment.**
562. Leaf roll of potato is caused by **Virus.**
563. Powdery mildews are controlled by spray of **Sulphur.**
564. Foot rot of Papaya is caused by ***Pythium aphanidermatum.***
565. First plant parasitic nematode discovered was ***Anguina tritici.***
566. Striga is a plant parasite included in **Phanerogamic plant parasites.**
567. Bacterial cell surrounded by flagella is called **Peritrichous.**
568. Transduction is a process of multiplication in **Bacteriophages.**
569. The organisms lacking a well defined nucleus are termed **Prokaryote.**
570. The organisms in the order Peronosporales cause the disease known as **Downy mildew.**
571. Dithane M-45 is a **Fungicide.**
572. Exclusion of plant disease by legislation is known as **Plant quarantine.**

573. In India Destructive insect and pest act was promulgated in the year **1914**.
574. Conidia and sporangia are the spores produced by **Fungi**.
575. Fungi which can grow only on living host plant are called **Obligate parasites.**
576. The genera Claviceps, Alternaria and Sclerospora are **Fungi**.
577. Disease symptoms known as phyllody in sesamum is caused by a **Mycoplasma like organisms.**
578. Nematodes are also known as **Tape worms.**
579. Plant parasitic nematodes possess **Three-part oesophagus.**
580. Father of American Nematology **N.A. Cobb.**
581. Which nematodes attack plant parts? ***Anguina tritici***.
582. The anal and vulval openings in nematodes are located on **Ventral.**
583. Most of the plant parasitic nematodes belong to order **Tylenchida.**
584. Sexual dimorphism means **Males and females are morphologically dissimilar.**
585. Who gave the term pasteurization? **Louis Pasteur.**
586. Tomato leaf curl is caused by **Virus.**
587. Term commensalisms means **Living together of the two species, one of which is benefited by the association whereas the other is neither beneficial nor harmful.**
588. Which are the cellulose attacking fungi at the time of decomposition of organic compound? ***Trichoderma, Aspergillus***.
589. Which is a poisonous mushroom? **Amanita.**
590. What shows the symbiotic relationship? **Rhizobium.**
591. Which fungus suppresses various pathogenic fungi present in soil? ***Trichoderma viridae***.
592. Indian Phytopathological society was established by **B.B. Mundkur.**
593. What is a gram positive bacterium? **Corynebacterium.**
594. Fusarium belong to order **Melanconiales.**
595. When bacterial cells are subjected to unfavourable conditions they form **Endospores.**
596. Association of fungi with roots of higher plants is known as **Mycorrhiza.**

597. Mycoplasma disease of plant was first reported by **Doi *et al* and Ishiie *et al*.**
598. The casual organism for earcockle disease of wheat is ***Anguina tritici*.**
599. Carbamate nematicide is **Aldicarb.**
600. Uncinula causes the disease **Powdery mildew.**
601. Powdery mildew fungi belong to the family **Erysiphaceae.**
602. First plant parasitic bacteria was reported by **T.J. Burill.**
603. Black heart of potato is caused due to deficiency of **Oxygen.**
604. In which rust pustules are arranged in end to end manner and forming stripe? **Yellow rust.**
605. Which is the root knot nematode? **Meloidogyne.**
606. Virus is transmitted by nematode **Xiphinema.**
607. Which is citrus nematode? ***Tylenchulus semipenetrans*.**
608. Molya disease of Wheat and barley is caused by ***Heterodera aveane*.**
609. Khaira disease of rice is caused by **Zinc deficiency.**
610. Smut gives the appearance of **Sooty powder.**
611. Nematodes penetrate in plant tissues with the aid of **Stylet.**
612. Who discovered the first antibiotics **Alexander Flemming.**
613. Which was the first plant bacterial disease reported? **Fire blight of apple.**
614. Which of the nutrient is supplied by mycorrhiza to the plant? **Phosphorus.**
615. Taphrina causes the disease **Peach leaf curl.**
616. Plant disease is written by (BIBLE) **R.S. Singh.**
617. Albugo is a **Obligate parasite.**
618. Primary source of infection to green ear disease of Bajra is **Seed and soil.**
619. Wheat rust life cycle in India was given by **K.C. Mehta.**
620. Brown spot of maize is caused by ***Physoderma maydis*.**
621. Most effective fungicide against powdery mildew disease **Karathane.**
622. Linseed rust is caused by ***Melampsora lini*.**
623. Solar energy treatment for the control of loose smut of Wheat was given by **Luthra and Sattar.**
624. Systemic fungicides was discovered by **Van schleming and Kulka.**

625. Heteroecious nature of rust means **A fungus require two different host species to complete life cycle.**
626. Smut of Wheat is effectively controlled by **Vitavax.**
627. Root parasite of Bajra is **Striga.**
628. In plains of India which wheat rust appear first **Yellow rust.**
629. Root rot of Papaya could be controlled by application of **Bordeaux mixture.**
630. Yellow vein mosaic of Bhendi was reported by **Uppal *et al.***
631. Leaf roll of potato is transmitted by ***Myzus persicae*, *Myzus preudosolani* and *Myzus circumflexum*.**
632. Black tip of mango is caused by **Born deficiency.**
633. *Tolyposporium ehrenbergii* cause **Long smut of sorghum.**
634. *Tolyposporium penicillariae* cause **Bajra smut.**
635. Whip smut of sugarcane is caused by ***Ustilago scitaminae*.**
636. Leveillula is a **Endoparasite.**
637. Downy mildew of maize is caused by ***Peronoscleropsora*.**
638. Tobacco mosaic virus was first reported by **Mayer.**
639. Stripe disease of barley is caused by ***Drechslera gramineum*.**
640. Alexander Fleming (1929) discovered **Penicillin.**
641. Death of tissue in virus infected plant is described as **Necrosis.**
642. Carborundum powder is **Silicon carbide.**
643. Secondary spread of ergot of bajra is through **Air and Insect.**
644. Indian journal of mycology and plant pathology is published from **R.C.A. Udaipur.**
645. Wilt of Arhar is caused by **Fusarium.**
646. Yellow ear rot of wheat is caused by **Anguina and Corynebacterium.**
647. Bengal famine was caused due to ***Drechslera oryzae*.**
648. The microorganism used in brewing industry is **Yeast.**
649. The plant disease caused by an algae is **Red rust of tea.**
650. A bacterium is devoid of **Chlorophyll and Well defined nucleus.**
651. A virus particle with both nucleic acid and protein coat is known as **Virion.**
652. Lichen is an association between **Algae and Fungi.**
653. Rusts belongs to sub division **Basidiomycotina.**

654. Sugar beet cyst nematode reported by **Schacht.**
655. Sporangium on germination produce vesicle in genera **Phytophthora.**
656. Phytophthora belongs to the family **Pythiaceae.**
657. Rhizopus belongs to class **Zygomycetes.**
658. Phyllactinia causes the disease **Powdery mildew.**
659. Tilletia causes the disease **Bunt.**
660. Ufra disease of Rice is caused by **Ditylenchus.**
661. Flag smut of wheat could be controlled by **Seed treatment with vitavax.**
662. Root knot of vegetables is caused by **Meloidogyne.**
663. Cysts are formed by **Heterodera.**
664. Meloidogyne belongs to group of **Endoparasite.**
665. Spermatheca is found in the nematode body of **Female.**
666. Didelphic means nematodes with **Two ovaries.**
667. Heterodera belongs to super family **Heteroderoidea.**
668. Trichodorus belongs to family **Trichodoridae.**
669. Mycology is the study of **Fungi.**
670. Cocci are **Spherical shaped.**
671. Who proved that virus could pass through filters with pores small enough to retain bacteria **Iwanowski.**
672. Gelatin plate technique was discovered by **Robert Koch.**
673. Botulism is caused by species of **Clostridium.**
674. Bacilli are **Rod shaped.**
675. Yellow pigment in the culture is produced by **Xanthomonas.**
676. Phycomycetes belong to **Fungi.**
677. Agar- Agar is produced from **Gelidium.**
678. In fermentation industry, Pasteurization is done at temperature **62.8°C.**
679. The sub-division ends with **mycotina.**
680. Acervuli are formed in order **Melanconiales.**
681. Loose smut of wheat can be controlled by **Vitavax.**
682. Fungicides which are absorbed into the system of plant and move to the remote site of infection are known as **Systemic.**
683. Zoospores of Oomycetes fungi are **Biflagellate.**

684. Choose the correct statement **Zoospores of chytridiomycetes and hyphochytridiomycetes are uniflagellate but tinsel type in former and whiplash type in later.**
685. Seed potato is not transported from Darjeeling area to other parts of the country because of **Domestic quarantine.**
686. The fungi-imperfecti are classified under the class **Deuteromycetes.**
687. Hermaphroditic fungi are those in which **Each thallus bears both male and female organs.**
688. White rust fungus, *Albugo candida* belongs to **Oomycetes.**
689. Which pathogen is associated with the discovery of Bordeaux mixture ***Plasmopora viticola*.**
690. Which pathogen is associated with Irish Famine? ***Phytophthora infestans*.**
691. Sexual spores in oomycetes are known as oospores while asexual spores as **Zoospores.**
692. Green ear disease (downy mildew) of bajra is caused by ***Sclerospora graminicola*.**
693. Damping off seedlings is caused by **Only *Pythium aphanidermatum*.**
694. Obligate organisms are those organisms which can obtain food from: **Only living protoplasm.**
695. Ascomycetes and Basidiomycetes are known as **Higher fungi.**
696. Peach leaf curl is caused by ***Taphrina deformans*.**
697. Which downy mildew fungus? ***Peronospora parasitica*.**
698. White rust of crucifers caused by *Albugo candida* belongs to class **Oomycetes.**
699. Viruses which can kill the bacteria are known as **Bacteriophages.**
700. The pathogenic organisms included under family Erysiphae cause **Powdery mildew.**
701. Powdery mildew of apple is caused by ***Podosphaera leucotricha*.**
702. Ascus normally contains definite number of ascospores which is **Eight.**
703. Ascomycetes produce their asci in fruiting bodies known as **Ascocarp.**
704. Sclerotia are the primary survival structures of ***Sclerotinia sclerotiorum*.**
705. Sesamum phyllody is caused by **Mycoplasma like organisms.**

706. Which one of the following diseases is internally seed borne **Loose smut of wheat.**
707. Dithane M-45 is available now as **Indofil M-45.**
708. Emisan is a **Organomercurial.**
709. Thiram is used as **Seed dresser.**
710. Karathane is a **Fungicide.**
711. When a disease is more or less constantly present from year to year in a moderate to severe form, in a particular country or part of the earth then it is known as **Pandemic.**
712. Who inverted Microscope? **Leeuwenhoek.**
713. First bacterial disease known as fire blight of pears was reported for the first time **Brefeld.**
714. The father of modern Plant Pathology is considered to be **E.J. Butler**
715. The abnormal increase in the size of a plant organ is known as **Hypertrophy.**
716. The term necrosis indicate **Death of cells.**
717. Dying of plant organs, especially stems or branches from tip to backward is known as **Die-back.**
718. Bacteria having one flagellum at each end of the cell are known as **Amphitrichous.**
719. Hyphal structures sent in to the host cell by different fungi to absorb the nutrients are known as **Haustoria.**
720. The most destructive phase of the bacterial blight disease of rice is known as **Kresek.**
721. Brown rot or wilt disease of potato is caused by (1st reported bacerial disease in india) ***(Rolstonia solanacearum)* bacterium.**
722. The tundu disease of wheat is usually associated with a nematode known as ***Anguina tritici.***
723. Stalk rot of maize is caused by ***Erwinia chrysanthemum pv. zea.***
724. Plasmodiophora brassicae causal agent of club root of crucifers was first fully described by **Woronin.**
725. Raising the pH of the field soil to 7 or above by adding lime (Calcium carbonate) gives good control of **Club root of crucifers**
726. Late blight of potato was introduced in India for the first time in **Nilgiri hills.**

727. Blight of Colocasia is caused by ***Phytophthora colocasia***.

728. Seedling blight of castor caused by Phytophthora parasitica was first recorded and studied by **Dastur**.

729. Koleroga disease of areca palms is also known as **Mahali disease**.

730. Systemic infection of *Albugo candida* in crucifers results in to **Hypertrophy**.

731. Green ear disease of bajra is also known as **Downy mildew**.

732. Downy mildew fungi are **Both seed and soil borne**.

733. Downy mildew of crucifers is caused by ***Peronospora parasitica***.

734. The source of primary inoculum of *Pseudoperonospora cubensis* is **Oospores**.

735. Which pathogen caused heavy losses to wine industry in France due to its epidemic in 1875? ***Plasmopora viticola***.

736. Resting structures of *Protomyces macrosporus* causing stem gall of coriander are **Ascospores**.

737. Taphrina deformans causes **Peach leaf curl**.

738. Leaf spot of turmeric is caused by ***Taphrina maculans***.

739. Characteristic symptoms of soft rot of apple appear on **Mature apples**.

740. Branching of sporangiophores of Plasmopora viticola to the main axis is at **Right angle**.

741. Secondary spread of powdery mildews is generally carried out by **Conidia**.

742. Cleistothecia are the primary survival structures of Erysiphe graminis which survive in the **Soil**.

743. Each cleistothecium of *Podosphaera leucotricha* contains **one ascus**.

744. Ring spot disease of sugarcane is caused by ***Leptosphaeria sacchari***.

745. The most common stage of the fungus causing stem rot of paddy is **Sclerotial**.

746. The sclerotia of ergot disease yield an alkaloid 'Ergotin' which when assimilated in the blood of man or cattle affects the **Nervous system**.

747. The asci of *Claviceps fusiformis* are cylindrical each of which contains 8 ascospores which are **Filiform and hyaline**.

748. The sphacelia stage of *Claviceps purpurea* is also known as honey dew stage which is full of **Conidia.**
749. Sclerotia of Claviceps purpurea germinate to produce **Perithecia.**
750. Major source of Primary inoculum of false smut of rice **Sclerotia.**
751. Apple scab disease which earlier was confined to Kashmir in India spread to Himachal Pradesh after **1973.**
752. The conidial cycle (Conidia-infection-conidia) of *Venturia inaequalis* is completed within **8-10 days.**
753. Apple scab fungus perpetuates through **Perithecia.**
754. Stem rot of Rozelle is caused by ***Sclerotium rolfsii.***
755. Organism of loose smut of wheat can be termed as **Internally seed borne.**
756. Solar energy treatment of wheat seed is recommended for the control of **Loose smut.**
757. Covered smut of barley is caused by ***Ustilago hordei.***
758. Smut of bajra is caused by ***Tolyposporium penicillariae.***
759. Karnal bunt of wheat was reported for the first time in 1931 by **Mitra.**
760. Paddy bunt disease is also known as **Kernel smut.**
761. Leaf smut of rice is caused by ***Entyloma oryzae.***
762. In which country was flag smut of wheat was observed for the first time **Australia.**
763. Stripe rust of wheat is also known as **Yellow rust.**
764. Which rusts is the earliest to appear on wheat in India? **Brown rust.**
765. Bean rust is caused by ***Uromyces phaseoli.***
766. Which of the combined seed treatments provide an effective control of stripe disease of barley? **Vitavax + Thiram.**
767. Causal agent of leaf spot disease of rice caused by *Helminthosporium oryzae* is also known as ***Drechslera oryzae.***
768. Kitazin and Hinosan are the fungicides which are used for the control of **Paddy blast.**
769. Anthracnose of bean is caused by *Colletotrichum lindemuthianum* ***Glomerella lindemuthianum.***
770. Fusarium wilt of banana is also known as **Panama disease.**
771. The fruiting bodies produced by the fungus *Ascochyta rabiei* are known as **Pycnidia.**

772. Charcoal rot of soybean is caused by **Fungus.**
773. Stem canker and black scurf of potato is caused by **One fungus.**
774. The activity of ***Rhizoctonia solani*** causing stem canker and black scurf of potato can be suppressed by a biocontrol agent **Trichoderma.**
775. Molya disease of barley, in India, was first reported from Rajasthan by Prasad *et a*l in the year **1959.**
776. Potato leaf roll virus belongs to **Luteovirus group.**
777. Which viruses transmitted through all the modes of the transmission *viz.* through sap, grafting, tubers and aphids **Vein-banding severe mosaic of potato.**
778. Yellow mosaic of legumes is caused by Mung bean yellow mosaic virus which belongs to **Geminivirus group.**
779. Mycoplasma like organisms as plant pathogens were proposed for the first time in 1967 by **Doi *et al*.**
780. Black tip or mango necrosis is caused by **Smoke and sulphur dioxide emitted from the brick kilos.**
781. The book entitled "Fungicides in Plant disease control" has been written by **Y.L. Nene and P.N. Thapliyal.**
782. Floral parts are converted into greenish leaf like structure in monocots is called **Green ear.**
783. Ragi mottle streak is transmitted by **Leaf hopper.**
784. Citrus tristeza is caused by **Virus.**
785. Moko disease of banana is caused by **Bacteria.**
786. Fragmentation of specialized hyphae gives **Oidia.**
787. An example for enveloped virus is **Tomato spotted wilt virus.**
788. The vegetative structure of the fungus which helps in the absorption of nutrients **Haustoria.**
789. Rhizoctonia root rot is common in **Alkaline soil.**
790. Monogenic resistance **Vertical resistance.**
791. Tyloses are produced after infection in **Xylem.**
792. Bunchy top of banana is transmitted by **Aphids.**
793. Aflatoxin is produced by ***Aspergillus flavus*.**
794. Characteristic leaf shredding symptoms of Downy mildew disease is found in **Jowar.**

795. Bacterial blight or Black arm of cotton is caused by **Bacteria.**
796. Anastomosis means fusion of **Vegetative cells.**
797. X-bodies are produced by **Virus.**
798. Virus reproduces by **Replication.**
799. The characteristic of a pathogen being able to cause disease is termed as **Pathogenecity.**
800. The external or internal alterations occurring due to disease is referred as **Symptom.**
801. A.P.S stands for **American Phytopathological society.**
802. The word fungicide has derived from language **Greek.**
803. Wart disease of potato caused by *Synchytrium endobioticum* **Endemic disease.**
804. The stem fungus *Puccinia graminis tritici* enters through **Stomata.**
805. A thick strand of somatic hyphae resembling a rod top is generally termed as **Stroma.**
806. A mat of hyphae giving rise to short conidiophores closely packed together is termed as **Synemma.**
807. Green ear is a proliferation of flower parts into greenish leaf like structure generally noticed in **Ragi and Bajra.**
808. Most of the virus diseases are more virulent when the temperature is in between **15 to 20°C.**
809. Potato dextrose agar do not contain **Beef extract.**
810. The mechanism involved in spore discharge is **Shooting.**
811. Fire blight on apple and pear by *Erwinia* bacterium was first reported by **Burill.**
812. The chain of events involved in disease development, including the stages of the pathogen and effect of disease on the host-this phenomenon is generally termed as **Disease cycle.**
813. A compound which prevents fungal growth without killing fungus is termed as **Fungistatic.**
814. Asexual spore of the basidiomycetes fungus is **Aeciospore.**
815. The fruiting body basidiocarp of basidiomycetes is produced by **Tertiary mycelium.**
816. Sterile threads arises at the base of the perithecium, upward direction between asci **Paraphysis.**

817. The fungal pathogen containing chitin in their cell wall in ***Claiceps purpurea***.
818. The sexual spore of *Pythium aphanidermatum* is **Oospore.**
819. The sexual spore of Rhizopus is **Zygospore.**
820. The sexual spore of *Gibberella fujikuroi* is **Ascospore.**
821. Vegetative spore of Fusarium is **Chlamydospore.**
822. Long distance moving spores in Rust **Teliospore.**
823. Clamp connection is most common in case of **Smuts.**
824. Clamp connection is most common in case of ***Ustilago tritici***.
825. The spores responsible for wheat rust in India is **Uredopsore.**
826. Dikaryotic spores, smooth born in chains in Rust fungi is **Teliospore.**
827. The multicelled spore in case of Rust is **Basidiospore.**
828. The dormant spore which is vegetative spore but not asexual spore is **Chlamydospore.**
829. Gametangial copulation mainly observed in ***Phytophthora* sp.**
830. In stinking smut of wheat, a fishy odour is due to **Trimethylamine.**
831. The sterilization temperature in autoclave for 15 minutes is **121°C and 15 psi.**
832. Complete root phanerogamic parasite **Orobanche.**
833. Complete stem parasite **Dodder.**
834. Partial root parasite **Striga.**
835. Partial stem parasite **Mistletoes.**
836. The slime molds comes under the class **Myxomycetes.**
837. The plasmodium or plasmodium like structure is seen in division **Myxomycota.**
838. The genus Olpidium and Physoderma belongs to the order **Chytridiales.**
839. Which division belongs to lower fungi **Myxomycota.**
840. Bread molds belongs to class **Zygomycetes.**
841. The mycorrhizal fungi producing spores singly in soil or on sporocarp containing zygospores, chlamydospore or sporangia belongs to class **Endogonales.**
842. The imperfect fungi wherein sexual reproduction and structures lacking/unknown comes under subdivision **Deuteromycotina.**

843. The rust and smut fungi belongs to the class **Hemibasidiomycetes.**
844. The club shape fungi is the special characteristic of subdivision **Basidiomycotina.**
845. The wood decaying and root rot fungi grouped under class **Hymenomycetes.**
846. Nucleic acid content of plant virus consists of **5 to 40%**.
847. Protein content of plant virus consists of **60 to 95%**.
848. The spherical shaped plant virus contain **Higher percentage of nucleic acid.**
849. Elongated shape plant virus contain **Lower percentage of nucleic acid.**
850. Nucleic acids of most of the plant virus consists of **ss RNA.**
851. In RNA viruses, thymine is replaced by **Uracil.**
852. The insects which feed on infected plants for several minutes / hours to few days before they accumulate the enough virus for transmission. These type of viruses are **Semi persistent.**
853. The stylet borne viruses persist in the vector for only few minutes to hours are known as **Non persistent.**
854. Viruses which persist in vectors for few days are known as **Persistent.**
855. Leaf hopper and plant hopper transmitted virus causes disturbances in plants that arise primarily in the region of **Phloem.**
856. Nepo viruses are **Grape fan leaf virus and Tobacco ring spot virus.**
857. Netu viruses are **Tobacco rattle virus and Pea early browning virus.**
858. The genomic content of Fiji virus is **ds RNA.**
859. The genomic content of Tobamovirus is **positive single stranded RNA.**
860. Use of mild strain for protection of plants against to severe strain of same virus is called **Cross protection.**
861. The process in which conversion of RNA to DNA is called **Reverse transcription.**
862. Circular nature of mini viroid was shown using **Electron microscope.**
863. Genome content of viroid is **ss RNA.**

864. Antibody is made up of only **Proteins.**
865. Antibodies are secreted by **lymphocyte.**
866. Antibodies are divided into **5 class.**
867. The measure of overall stability of the complex between antibodies and antigens is called **Affinity.**
868. The measure of strength of binding of epitope to an antibody is called **Avidity.**
869. The relative measure of the concentration of specific antibody in an antiserum is **Titer.**
870. Mathew's plant virology is written by **R. Hull.**
871. The density gradient centrifugation as a method for purifying virus was first developed by **Brakke, 1951.**
872. A virus having multipartite **Alfalfa mosaic virus.**
873. The genome size of potyvirus is **8.5-10 kb.**
874. A citrus tristeza virus belongs to **Closteroviridae.**
875. Papaya ring spot virus transmitted by several aphid spp in the **Non persistent.**
876. Zucchini yellow mosaic virus is transmitted by **Aphid.**
877. The term clostero viruses means **Thread like viruses.**
878. Grassy shoot of sugarcane is transmitted by **Leaf hoppers.**
879. The parenchyma close to the veins in green and rest of the lamina surface shows chlorosis is called as **Vein banding.**

OOO

CHAPTER 8

Horticulture

1. National GDP – **5% from horticulture.**
2. Fruits and vegetables occumpy **65% area under cultivation.**
3. Indians spend about **50% of household expenditure** on food items.
4. **Temperate-zone fruits** (apples, apricots, cherries, and peaches) and walnuts.
5. Fruit rotting can be checked by slowing down the action of enzyme polygalactouronse of the organelle- **Glyoxysome.**
6. **Asafoetida (Devil's Dung)** is dried latex (oleo-gum-resin) exuded from the living rhizome.
7. **More than 50 types of spices with annual production of 3.02 million tonnes,** India is the largest producer, consumer and exporter of spices.
8. **With above 9500 spices from medicinal and aromatic plants,** India is accounting for 25-30 per cent of the world's production.
9. **India home to 49000 plant species** equating to about 12% of the known species in the world.
10. Sinigrin a glycoside is found in- **Black mustard.**
11. **A fruit**- Fennel, Caraway, Black pepper, Piper.
12. Soft seeded variety of Mango – **Sindhu.**
13. Mango variety suitable for high density planting- **amrapalli.**
14. IARI released variety **Mallika** is a cross between-**Neelam x Dasheri.**
15. IARI released variety **Amrapalli** is a cross between-**Dasheri x Neelam.**
16. IARI released variety **Ratna** is a cross between- **Neelam x Alphonso.**

17. IARI released variety **Sindhu** is a cross between-**Alphonso x Neelam.**
18. Poor man's root crop- **Tapioca.**
19. Propogation method in pineapple- **Suckers.**
20. Priyadarshini is a variety of – **Rose.**
21. Temperate crops-**Apple, Cabbage, Cauliflower, Carrot, Potato.**
22. Poor man's and arid fruit- **Ber.**
23. Poor man's root crop- **Tapioca (Cassava).**
24. "Queen of beverage crops" also Calcifuge crop- **Tea.**
25. "Kalpavariksha"-**Coconut.**
26. King of temperate fruits - **Apple.**
27. Queen of Spices – **Small cardamom.**
28. King of fruit – **Mango also called Bathroom fruit.**
29. Spice of the angel – **Fennel.**
30. An appetizing Vegetable- **Brinjal.**
31. An evergreen fruit tree- **Sapota.**
32. Brown rot of cauliflower- **B deficiency.**
33. W-2 of Papaya – **Highest content of papain.**
34. Onion is usually grown as –**late Kharif but in Maharathsra in Rabi.**
35. Umbel inflorescence is seen in- **Coriander.**
36. Water melon without seeds- **Triploid.**
37. Non volatile pungent oil obtained from Ginger-**Gingerly.**
38. **Major pigment in turmeric** – Curcumin.
39. Pungent principle of black pepper- **pipeline.**
40. Toxic alkaloid in dioscorea- **Discoing.**
41. Saponin produced by yam-**Sapegemin.**
42. Plants having modified rhizomes adapted for storage of food - **Tubers.**
43. **Saffronin is extracted from** - Only from corolla.
44. **Rubber is chemically** Polyisorprene.
45. **Seedless fruits are developed as a result of** - Parthenocarpy, Apomixis, Parthenogenesis.
46. **Variety of beet is** Detroit dark red.
47. Priyadarshini is a variety of - **Rose.**

48. Delight and Perlette are promising cultivars of - **Guava.**
49. Pusa Snowball a late cauliflower variety developed by **IARI.**
50. National fruit of India, King & Pride, Choice and symbol of India- **Mango.**
51. 42% fruit area under- mango and almost all Indian varieties are **monoembryonic.**
52. In rain fed crop yield of dry chillies is **5-10 q/ha.**
53. KashiAmrit (DVRT-1) tomato variety developed by – **IIVR.**
54. **CAM plant**- Pineapple.
55. First Spices Park in Rajasthan **for processing of seed spices like cumin, coriander and fennel in Jodhpur, Rajasthan.**
56. India's export of seed quality items in 2010-11 was of value USD 101 million. India's share in world trade of seed items is **1.22%**.
57. **A flower**-Cloves, Saffron.
58. Multi layered cropping is practiced in – **Karnataka and Kerala.**
59. In budding and grafting the union between stock and scion takes place through –**Cambium.**
60. Russeting in apple is due to- **High temperature.**
61. Cell division continues till maturity in fruit development of – **Avocado.**
62. A chemical most effective in breaking bud dormancy in grapes- **Hydrogen cynamide.**
63. A chemical used for artificial ripening in banana- **Ethephone.**
64. Application of which chemical acts as an aid for mechanical harvesting- **ABA.**
65. High cytokinin/ auxin ratio in callus favours- **Shoot formation.**
66. Sprays of which of the following induces femaleness in fruits- **Maleic hydrazide.**
67. Maleness in fruits can be induced by application of – **GA_3.**
68. Fruit thinning growth regulator is- **NAA.**
69. Fruit on the basis of respiration can be classified as - **Climactric and nonclimactric.**
70. Losses during transit are more in- **Banana.**
71. Pineapple is highly suited for regions- **Topical humid.**
72. Bt gene is incorporated for – **Insect resistance.**

73. Botanically not a fruit- **Cashew apple.**
74. Maximum heating is obtained with which mulch material- **Black.**
75. D leaf is a best indicator of nutrient status of- **Pineapple.**
76. Features of cold desert – **Heavy influx of IR & UV light, Low air density.**
77. Highly shade tolerant fruit crop is- **Carambola.**
78. Highly drought tolerant fruit crop is- **Ber.**
79. Fruit crop coming well in acidic soil is – **Srawberry, Raspberry, Gooseberry.**
80. Well decomposed manure should have C:N raio – **10:1.**
81. Best time for orchard soil culture is – **November - January.**
82. Soil erosion in orchard can be checked by – **Sod culture.**
83. Horticultural corps though cultivated on 5% of cropped area of country, consume ----% of total pesticides used in India-**18.**
84. Spices export from India during 2008-09 is worth – **Rs. 5000 crore**
85. Our major spice importer is – **USA.**
86. The export destination of ginger from NE states is- **Bangladesh.**
87. Major markets of Indian spice oil and oleoresins products is-. **USA.**
88. The present area under fruit crops in the country is ------- thousand ha -. **5510.**
89. Productivity of fruit crops in India is ---- t/ha- **10.7.**
90. Productivity of vegetable crops in India is ----- t/ha- **15.2.**
91. Cropwise maximum productivity is obtained in which vegetable- **Tapioca.**
92. The word horticulture is derived from Hortus and Coleure meaning garden and cultivation. These words are- **Latin.**
93. National Horticulture Board was established in – **1984.**
94. IIHR Banglore was established in -**1968.**
95. Pineapple and banana are – **Ornithophilous.**
96. Quincunx system of planting accommodates how much more plants than square system of planting-**89%.**
97. HDP in mango follows which system of planting- **Triangular.**
98. Commercial method of propagation of acid lime is through- **Seed.**
99. Banana is propagated through- **Sword sucker.**
100. Banana suffers heavily if temperature falls below- **10°C.**

Spices	Fruit type	Major chemical constituent
Ajowan	Seed	Thymol
Anissed	Fruit	Anethole, anisaldehyde
Bay leaf	Leaf	Linalool, methyl eugenol
Cardamom	Fruit	1, 8-cineole,
Celery	Fruit	Myrcene, limonene
Cinnamon	Bark	Eugenol, benzyl benzoate, cinnamaldehyde
Clove	Buds	Eugenol, Eugenyl acetate
Garlic	Bulb	Curcumin
Ginger	Rhizome	Gingerol, zingiberene

Fruit types

Berries	Banana, Brinjal, Phalsa, Tomato, Grapes, Chillies, Guava, Sapota
Drupe	Mango, Peach, Plum, Cherry, Apricot, Coconut
Hesperidium	Citrus
Amphisarea	Wood apple, bel
Balusta	Pomegranate
Sorosis	Pineapple
Syconium	Fig
Nut	Cashew
Peppo	Cucurbits
Hips	Rose
Capsule	Okra

Methods of propagation

Mango	Epicotyl grafting, veneer grafting
Banana	Sword suckers
Citrus	Shield budding, T Budding
Grapes	Hard wood cutting
Litchi	Air layering
Guava	Stooling
Sapota	Inarching
Strawberry	Ruman
Pineapple	Suckers
Sweet potato	Root cuttings, vine cutting

- **20 million people** engaged in production, processing, transportation and marketing of betel leaves.
- **Betel leaves** are being cultivated in Andhra Pradesh, Assam, Bihar, Karnataka, Kerala, Madhya Pradesh, Maharasthtra, Odisha and West Bengal.
- **Vegetable Initiatives** for Urban Cluster (VIUC) has been started
- **Presently, coconut is** cultivated over 18.95 lakh with estimated production of 15730 million nuts and productivity of 8303 nuts per ha in the country. India stands 1st in productivity and 2nd in production of coconut nuts in the world.

Fruit disorders

Sunscald	Pineapple
Multiple crown	Pineapple
Water berries	Grapes
Shot berries	Grapes
Coloure	Grapes
Water berries	Grapes
Crown choking in coconut	B deficiency
Bud and leaf malformation	B deficiency
Barrenness in coconut	B deficiency
Mahali and yellow leaf disease	Arecanut
Buttoning	Cauliflower
Fruit necrosis of "Aonla"	B deficiency

- Queen of beverage and calcifuge crops- **Tea**.
- Kalpavariksha- **Coconut**.
- Flat garden, Hill garden examples of- **Japanese garden**.
- Features of Japanese garden are- **Lanterns, Pagodas, Bridges**.
- **Root Crop** - Yam, sweet potato, tapioca.
- **Cucurbits** - Snake gourd, Pumpkin, Cucumber.
- Vertical mulching is mostly practiced in --- **Coffee garden**.
- A rhizome- **Turmeric, Ginger**.
- Allicin in large amounts is mostly found in- **Turmeric**.
- Bleeding heart- ***Dicentra***.

- Horse radish- ***Armoracia rusticana*.**
- Crops require seed beds for sowing – **Onion, Lucerne and finger millet.**
- Pusa Snowball a late cauliflower variety developed by **IARI.**
- Kashi Amrit (DVRT-1) tomato variety developed by – **IIVR.**

Insect pest resistant varieties

Crop	**Pest**	**Varieties**
Brinjal	Shoot and fruit borer Jassids Nematode	Doli 5, Punjab JamuniGola, Punjab Sadbahar, Sm 17-4, PBr 129-5, Punjab Barsati, ARV 2-C, Pusa Purple Round, Punjab Neelam, Kdalyanpur-2 and Punjab Chamkila. Black Beauty, Ram Nagar Giant and Vijay Hybrid. Pusa Purple Cluster, Pant-Rituraj, Arka Nidhi, Arka Kesav & Arka Neelkanth.
Cabbage	Aphid	All Season, Red Drum Head, Sure Head, and Express Mail.
Cauliflower	Stem borer	Early Patna, EMS-3, K W- 5, K W-8 and Kathmandu Local.
Okra	Jassid	IC-7194, IC-139999, New Selection and Punjab Padmani.
Round gourd	Fruit fly	Arka Tinda
Pumpkin	Fruit fly	Arka Suryamukhi
Bitter gourd	Fruit fly	Hissar-II

- Fruits and vegetables, together, constitute about **92%** of the total horticultural production in the country.
- India is the **second largest** producer of fruits in the world, it is the **largest producer** of fruits like mango, banana, papaya, sapota, pomegranate and aonla.
- Fruits share **32%** in total production.

Area under Horticulture Crops in 2011-12	22.25 million ha
Production under Horticulture Crops in 2011-12	247.54 million tonnes
Productivity under Horticulture Crops in 2011-12	11.13 tonnes per ha
Area under Fruit Crops	6.58 million ha
Production of Fruit Crops	77.52 million tonnes
Area under Vegetables Crops in 2011-12	8.49 million ha
Production under Vegetables Crops in 2011-12	149.61 million tonnes
Productivity under Vegetables Crops in 2011-12	17.42 tonnes per ha.
Floriculture During 2011-12 Covered Production	0.19 million ha
Production of Loose Flowers	1.03 million tonnes of
Production of Cut Flowers	69,027 million numbers
Area under Spices	3.03 million ha.
Production of Spices	5.73 million tonnes from
Area under Plantation Crops	3.35 million ha
Production under Planatation Crops	12.99 million tonnes

- **Indian basilowes** its strong pungent taste to a sesquiterpenoid - caryophyllen and a phenylpropanoid, methyl eugenol (both around 30%) plus minor amounts of methylchavicol (10%).
- The bay leaves have essential oil from the leaves containing **cinnamic aldehyde** and traces of eugenol as its main constituents.
- Richest source of vit. C – **Barbadose cherry.**
- Protective foods- **Fruits & Vegetables.**
- Self incompatibility phenomenon used for – **seed production in cabbage.**
- Most utilized part of **Bishop's Weed** is the small, caraway like fruits that have the essential oil (2.5 to 5% in the dried fruits) being dominated by thymol (35 to 60%); a- pinene, p- cymene, limonene and terpinene as its main constituents.
- B.P. Pal variety- **rose.**
- **Cassia contains** 4% essential oil of which 75 to 90% are composed by cinnamic aldehyde and eygenol is found in traces along with small amounts of coumarin.
- **Chillies** have pigment **Capsanthin** and pungency attributed by **capsaicin.**

- Finest quality ginger is grown in Kerela. Indian dry ginger is known in the world market as **'Cochin Ginger' (NUGV) & 'Calicut Ginger' (NUGK).**
- **Biocontrol agents** *–Trichogramma brassiliensis-Bacillus thuringensis, Okra shoot and fruit borer.*
- **Diamond back moth-** Shoot and fruit borer of okra and tomato fruit borer.
- Putrefication of meat, eggs & souring of milk is due to – **Bacteria.**
- Saffron spice use – **Stigma.**
- Export growth of fresh fruits and vegetables in term of value is 14% and of processed fruits and vegetables is **16.27%**.

Horticulture in INDIA –State ranks

Commondities	State ranks in production
Apple	JK>Himachal>Uttranchal>Arunachal Pradesh
Cashew	Kerala>MH>TN (area) and MH>Kerala>AP (production)
Coconut	Kerala>Karnataka>TN (area) and Kerala>TN>Karnataka (production)
Banana	Maharashtra>TN>Gujarat>Karnataka
Grapes	Maharashtra>Karnataka>TN>AP
Guava	UP>MP>Punjab>Bihar
Lichi	Bihar>WB>Jharkhand>Assam
Mango	Ap>UP>Karnataka>Bihar
Papays	Ap>Gujarat>WB>Karnataka
Pineapple	WB>Assam>Karnataka>Tripura
Pomegranate	Maharashtra>Karnataka>AP>Gujarat
Citrus	AP>Maharashtra>Punjab>MP
Sapota	Maharashtra>Gujarat>TN>AP
Coffee	Karnataka>TN>Assam>Kerala
Jute	WB>Bihar>Assam>AP

Abbreviations:

TN = Tamil Nadu, AP = Andra Pradesh, MP = Madhya Pradesh, WB = West Bangal

101. **Common preservative and their uses** – Preservative, Salt, Nitrates, Sulfur Dioxide and Sulphites, Benzoic Acid and Sodium Benzoate, Propionic acid propionates, Sorbic acid and sorbates, Sodium diacetate, Lactic acid, Sugar, Wood smoke.

Varieties – fruits

Mango	Mallika, Amrapali, Ratna, Sindhu, Arka Puneet, Dashehari-51 hybrids namely CISH-M-1 and CISH-M-2
Guava	Lalit, CISH-G-1 and CISH-G-2, . L-49
Banana	FHIA-01 and FHIA-03, Panchananda, Bluggoe, Saba, Col H1 and H2
Grape,	Perlette Beauty Seedless and Pusa Seedless Anab-e-Shahi, Dilkhush, Thompson Seedless, Tas-A-Ganesh, Sonaka, Banglore Blue and Panchadraksha
Citrus	Rough Lemon,Pramaini,Vikram PKM-1, Rangpur Lime, Flying Dragon and Rich 16-6
Lemon	Gandhraj
Papaya	Co-1 to Co-7, Coorg Honey Dew, Pusa Delicious, Pusa Majesty, Pusa Giant, Lal Ambari, Sunehari. Red Spur, Star Crimson, Golden Spur, Red Chief, Oregon Spur, Skyline Supreme and Vance Delicious

Traditional regions of rubber cultivation in (Kerala and Kanyakumari district of Tamil Nadu) and not- traditional regions other than North East (mainly Karnataka, Maharashtra, Andhra Pradesh, Goa, Odisha and West Bengal.

1. Consumption of natural rubber in India as **966,750 tonnes in 2011-12**.
2. Bleeding heart- ***Dicentra***.
3. The major protein found in silk is-**Fibroin**.
4. Horticultural crops though cultivated on 5% of cropped area of country, consume -----% of total pesticides used in India- **18**.
5. Spices export from India during 2008-09 is woth – **Rs. 5000 crore**.
6. Our major spice importer is – **USA**.
7. The export destination of ginger from NE states is- **Bangladesh**.
8. Major markets of Indian spice oil and oleoresisns products is-. **USA**.
9. The present area under fruit crops in the country is ----- thousand ha -. **5510**.
10. Productivity of fruit crops in India is ----- t/ha- **10.7**.
11. Productivity of vegetable crops in India is ----- t/ha- **15.2**.
12. Fruit crops production in country share is maximum – **Banana**.
13. Cropwise maximum productivity is obtained in which vegetable- **Tapioca**.

14. The word **horticulture** is derived from Hortus and Coleure meaning garden and cultivation. These words are- **Latin.**
15. Quincunx system of planting accommodates how much more plants than square system of planting-**89%**.
16. HDP in mango follows which system of planting- **Triangular.**
17. Coomercial method of propagation of acid lime is through- **Seed.**
18. Banana is propagated through- **Sword sucker.**
19. Banana suffers heavily if temperature falls below- **10°C.**
20. ArkaPuneet is a cross between - **Alphonso x Banganpalli.**
21. ArkaAruna is a cross between - **Banganpalli x Alphonso.**
22. Richest source of fat- **Walnut.**
23. **Banana** –Shows double sigmoid growth curve.
24. Desuckering is followed in – **Banana.**
25. Giving shape to the plant is called as - **Topping.**
26. Virachola Isocrates is the fruit pest of - **Pomegranate.**
27. Monoembryonic species of citrus- ***Tahiti lime***.
28. Monoembryonic species of citrus- **Citron.**
29. "Degreening" of Nagpur mandarin is done with - **Ethylene.**
30. Used as an indicator plant for Tristeza - **Kagzi lime.**
31. Sweet lime –***Citrus limmettoides***.
32. Acid lime –***Citrus aurantiflia***.
33. Lemon - ***Citrus limon***.
34. Granulation is characterized by increase in- **Pectin, Lignin, Polysaccharide.**
35. Fruit necrosis of "Aomla" is due to-**Boron deficiency.**
36. 'Bahar' treatment in guava by- **NAA, Ethepon, Urea.**
37. Marmalade is a processed fruit product of- **Citurs.**
38. Flat garden is an example of - **Japanese garden.**
39. Features of Japanese garden - **Lanterns, Pagodas, Bridges.**
40. Mango malformation can be checked by - **NAA.**
41. Induce male sterility in grapes- **Maleic hydrazide.**
42. Coloure, Shot berries, Water berries are disorders in- **Grapes.**
43. Sunscald, Multiple crown are disorders in - **Pineapple.**
44. Barrenness, Crown choking, Bud and leaf malformation in coconut - **Boron deficiency.**

45. Bagging of anar fruits with polythene cover is to avoid damage from : **Butterfly sucking moth.**
46. The infested region on bark turns into pink colour and pink colour encrustation on apple fruits are the characteristic damage symptoms of : **San jose scale, *Quadraspidiotus perniciosus*.**
47. Egg- larval parasitoid, Chilonusblackburni is used to check the population of : **Potato tuber moth, *Gnorismoschema/Phthorimaea operculella*.**
48. Minning and skeletanization of cabbage leaves are the characteristic damage symptoms of - **Diamond backmoth, Plutellaxylostella.**
49. Bt cabbage is developed against - **Diamond backmoth (DBM).**
50. Gummosis and crown toppling in coconut are due to - **Red Palm Weevil.**
51. An introduced pest of coffee - **Berry borer.**
52. In storage, grains can be mixed with - **Malathion dust (5%).**
53. **Long Day Fruit** - Passion Fruit, Apple.
54. **Short Day Fruit** - Strawberry, Pineapple, Coffee.
55. **Day Neutral** - Papaya, Guava, Banana.
56. **Cucurbits** - Snake Gourd, Pumpkin, Cucumber.
57. **A Root Crop** - Yam, Sweet Potato, Tapioca.
58. Total Root Parasite - **Orobanche.**
59. Semi Root Parasite - **Striga.**
60. Total Stem Parasite - **Cuscuta.**
61. Orobranche is also known as **Broom Rape.**
62. Vertical Mulching is Mostly Practices in - **Coffee Garden.**
63. **A Vegetable** - Okra, Chilli, Brinjal, Tomato.
64. **Tree for Forage and Construction**- Teak, Tree Bean, Saal, *Terminalia.*
65. Chinese Evergreen - ***Aglaonema*.**
66. **Sleeping Sickness**- Brush Weed.
67. **In Mint which Weed Reduces the Quality of the Oil**- Cirsium
68. Potato Virus is **- Aphid.**
69. Tomato Leaf Insect is - **White Fly.**
70. Bunch Top of Papaya - **Leaf Hopper.**

71. Bunch Top of Banana - **Aphid.**
72. Furrow Irrigation Method is Suitable for - Cotton and Vegetables.
73. Pusa 1080, 1103 and 1105 Related - **Chickpea.**
74. Pusha Surya and Pusa Arunima - **mango.**
75. Pusaratna is variety of - **mungbean.**
76. Climactic fruit- **mango, apple, kiwi, melon, guava, papaya, banana, peach, pear**.
77. **Non** climactic fruit- **citrus, litchi, strawberry, pomegranate, grapes, loquat**.
78. Giving shape to the plant is called as --- **topping.**
79. Noxius weed of south India- **Wild brinjal.**
80. Currently in India the processing level is estimated to be around 2.20%.
81. Heading back done within a desired distance of - **15-25 cm.**
82. Thinning out removes – **Shoot, Spur, Branch.**
83. Open system of training is common in- **Pear.**
84. **Effect of pruning is** – Delayed bearing, Prolonged vegetative growth, Check in formation of flower and fruit bud.
85. Commercial method of grape propagation is – **Hard wood cutting**
86. Ultra dwarf rootstock of apple is **M27.**
87. Rootstock of citrus for shallow soil and high rainfall area – **Trifoliate Orange.**
88. Salt creek and Dogridge are rootstocks for- **Grape.**
89. Manilkarahexandra is commercial rootstock for- **Sapota.**
90. Is method of crop regulation- **Pruning, Smudging, Withholding irrigation**.
91. Strawberry is resitant to deficiency of which element- **Mn.**
92. Green veins on a light yellow to green background symptom of- **Fe deficiency.**
93. Little leaf of mango is caused by deficiency of- **Zn.**
94. Die back of citrus can be attributed to deficiency of- **Cu.**
95. GA has effect on rooting- **Antagonistic.**
96. BOA hastens flower initiation in – **Papaya.**
97. Pineapple flowering is induced by- **Ethrel.**
98. Fruit drop in citrus can be controlled by-**2,4-D.**

99. **Alternate bearing is** common in - Mango, Apple, Olive.
100. Mango malformation is not observed in – **South India.**
101. Regular bearing variety of mango is- **Neelam.**
102. Improper finger filling in banana is due to deficiency of – **K.**
103. Cork spot of apple is due to- **Water deficiency.**
104. Hen and chicken is a disorder of – **Grape.**
105. Black heart in apple is caused due to – **Low temperature.**
106. Spongy tissue is disorder of- **Alphanso.**
107. **Citrus decline symptom** is- Death of twig, Reduction in leaf size and twig.
108. Internal fruit necrosis in aonla is due to- **B deficiency.**
109. Exanthema in citrus is due to deficiency of- **Cu.**
110. Shot berry of grape can be controlled by- **Ethephone**.
111. **Fascination of pineapple is** due to- High fertile soil, Nutritional imbalance.
112. Totipotency of plant cell was proposed by- **Haberlandt.**
113. In India, commercial exploitation of micro propagation is limited to- **Banana.**
114. For berry elongation in grape which compound is used-**Gibberelin.**
115. Richest source of fat is – **Walnut.**
116. Bael is the richest source of- **Riboflavin.**
117. Ascorbic acid is highest in – **Barbadose cherry**.
118. Mattocking is associated with- **Banana.**
119. Guava wilt is common in pH range of-**7.5–9.0.**
120. Stenospermocarpy in mango is observed in- **Sindhu.**
121. Maximum papain yieliding papaya selection is – **°C–6.**
122. Hesperidium is botanical form of fruit in- **Citus.**
123. Strawberry is propagated through- **Runner.**
124. Vegetative propagation of date palm is through – **Sucker.**
125. Best planting material of pineapple is- **Slip.**
126. Fruit is not monocotyledonous-**Pineapple.**
127. Citrus type is monoembryonic—**Pummelo.**
128. Kaverian improved variety of Passion fruit was developed at IIHR, Bangalore. The parents of Kaveri are- **Purple and Yellow.**

129. Mauritius a popular variety of pineapple is classified as- **Mid season.**
130. Degreening in citrus is done with the application of – **Ethephone.**
131. Kagzi lime is the best indicator for – **Tristiza.**
132. Bahar treatment in Citrus is practiced in- **Central and West India.**
133. Fruits are summer dormant/ deciduous- **Ber, Pomegranate, Bael.**
134. Bronzing in guava is associated with the deficiency of – **Zn.**
135. **Indigenous to India** – Mango, Banana, Bael, Jackfruit, Aonla, Phalsa.
136. Stone fruit formation is a disorder of – **Annona.**
137. Two leaf and a bud is type of plucking in – **Tea.**
138. Theobromin in cacao is extracted from- **Bark.**
139. Mango variety most suitable for high density planting is- **Amrapali.**
140. Somatic chromosome number of mango is—**40.**
141. Modified stem of banana is called-**Rhizome.**
142. Storage temperature of banana is------ °C–**13.**
143. Flower bud differentiation in mango in north India takes place in-. **Oct-Nov.**
144. Centre of origin of guava is- **Chile to peru.**
145. Optimum temperature for stratification of apple seed is ---- °C–**4.**
146. Fruits is the richest source of Vit C and can be used to avoid scurvy - **Aomla.**
147. Fruit crop suitable for cold desert is – **Apricot.**
148. Growth regulator acts anti to gibberelin – **Cytokinin.**
149. Chief active ingredient of the medicinal value of bael is- **Marmelosin.**
150. Which portion of plant is considered free from virus – **Meristem.**
151. Pigment responsible for colouration in papaya-**Caricasanthin.**
152. Litchi fruit is botanically – **Single seeded nut.**
153. Ploidy level of all cultivated strawberry is - **Octaploid.**
154. Edible portion of pineapple is - **Bract & perinath.**
155. HDP in pineapple accommodates approx. Plants/ ha-**60000.**
156. Is most suitable for intercropping in fruit orchard-**Short duration fruit plants.**

157. Most widely used plant growth substance in viticulture is-**GA_3**.
158. Chemical formula of potassium metabisulphite is-**$K_2S_2O_7$**.
159. Sodium benzoate is added for preservation of fruit juice in concentration of - **%-0.06-0.1.**
160. Curling of leaves in chillies is caused due to - **Thrips.**
161. Harvest of mangoes is more difficult due to presence of- **Mealy bugs.**
162. A regular pest of Okra is - **Fruit borer.**
163. Katte disease in groundnut is due to- ***Cercospora personata and C. arachidicola.***
164. A major pest of tomato is-**Fruit borer (*Helicoverpa armigera*).**
165. Stem girdler is a pest of - **Grape.**
166. Major pest of sapota - **Seed borer.**
167. Fruit sucking moth can be controlled by-**Bagging of fruits.**
168. Black fly is a serious pest of - **Citrus.**
169. Major pest of cashew is - **Tea mosquito bag.**
170. Cucumbers major pest is-***Bactrocera cucurbitae*.**
171. Premature fruit drop in chillies is prevented by-**NAA.**
172. Mango inflorescence contains which type of flower **–Male and hermaphrodite.**
173. Cultivated autopolyploid - **Potato.**
174. Ranks first in terms of processed/canned vegetable-**Tomato.**
175. Red colour of tomato-**Lycopene**, Yellow-**Carotnoid**, Orange-**Pic-lycopene.**

OOO

CHAPTER 9

Agricultural Economics

Microeconomics

1. Micro economics is also called as ________ **Price theory.**
2. Wealth definition of economics given by ________ **Adam Smith.**
3. Father of economics ________ **Adam Smith.**
4. Welfare definition of economics given by ________ **Alfred Marshall.**
5. Scarcity definition of economics given by ________ **Lionel Robbinson.**
6. Growth definition of economics given by ________ **Samuelson.**
7. Types of goods based on transferability given by ________ **Marshall.**
8. Example for Non-transferable external material good ________ **Degree certificate.**
9. Example for Non-transferable internal good ________ **Personality.**
10. Example for Non-transferable external personal good ________ **Friendship.**
11. Giffen goods are viewed from the angle of ________ **Income.**
12. The concept of General Equilibrium was given by ________ **Walras.**
13. LDMU concept given by ________ **H.H. Gossen.**
14. Total utility is maximum when marginal utility is equal to ________ **Zero.**
15. Processing of good gives ________ utility ________ **Form utility.**

16. Transportation of goods gives __________ **Place utility.**
17. Storage of good gives __________ **Time utility.**
18. Canons of Taxation was given by __________ **Adam Smith.**
19. In case of complementary goods the cross demand curve is __________ **Zero.**
20. Right shift in demand curve is due to __________ **Rise in price of its substitutes.**
21. Left shift in demand is due to __________ **Decrease in price of its substitutes.**
22. Upward movement along a demand curve is due to __________ **Rise in price.**
23. Downward movement along a demand curve is due to __________ **Fall in price.**
24. Rightward shift in supply curve is due to __________ **Rise in price.**
25. Leftward shift in supply curve is due to __________ **Fall in price.**
26. The goods X & Y are independent if cross elasticity of demand is __________ **Positive.**
27. The goods X & Y are substitutes if cross elasticity of demand is __________ **Zero.**
28. Demand for luxuries is __________ **Elastic.**
29. Consumer surplus = __________ **Demand price - actual price.**
30. Producer surplus = __________ **Market price – supply price.**
31. If the demand is unitary (EP=1) then the Marginal Revenue curve (MR) will be **Inelastic.**
32. If the demand is inelastic then MR will be __________ **Positive.**
33. Cross elasticity of demand between petrol and automobile is __________ **Zero.**
34. When there is an increase in demand of 'X if the price of 'Y' increases, X & Y are __________ **Substitutes.**
35. Downward movement along the demand curve indicates __________ **Contraction.**
36. Which type of tax helps in reducing disparities of income __________ **Progressive tax.**
37. Corporate tax is imposed by __________ **Central Government.**

Macroeconomics:

38. Macro economics also known as __________ **Income Theory.**
39. Return on saving is called as __________ **Interest.**
40. Change in capital stock is called as __________ **Investment.**
41. Keynes theory is a theory of __________ **short run.**
42. Theory of Rent __________ **David Ricardo.**
43. Quasi Rent __________ **Alfred Marshall.**
44. Subsistence Theory of Wages __________ **David Ricardo.**
45. Wage Fund Theory __________ **J.S. Mill.**
46. Residual Claimant Theory __________ **R.A. Walker.**
47. Time Principle Theory of Interest __________ **Irving Fisher.**
48. Migration and Unemployment Theory __________ **Harris and Todaro.**

Production Economics and Farm Management:

49. Indifference curve is also known as __________ **Iso utility curve.**
50. Competitive nature of human wants has given us __________ **Law of substitution.**
51. Manufacturer use both machinery and labour, then wants are __________ **Complimentary.**
52. Farm business income is __________ **GR-cost A1.**
53. Operating cost ratio is __________ **variable cost/ gross return.**
54. Possession of livestock is __________ **intermediate asset.**
55. Which method of depreciation is useful in case of livestock in early years of life __________ **annual revaluation method.**
56. Which of the following system of farming operationship is collective __________ **co-operative joint farming.**
57. Production for only home consumption is __________ **subsistence farming.**
58. bi> 1 represents __________ **increasing returns to scale.**
59. Which line represents isoclines __________ **Ridge line.**
60. When Y1 /Y2 is positive for substitution of 2 products Y1 and Y2 then relation is __________ **complementary.**
61. PPC for greater level of production represented by __________ **Moving away from origin.**

62. **Production possibility curve is also known as __________ iso-factor curve.**
63. **At the point of convergence of isoclines __________ Output is maximum, MR=0.**
64. **If factors are perfect complements then shape of the isoquant are __________ point.**
65. **Given that Y/X =10 and elasticity = 0.40, what is the value of marginal physical product __________ 4.00.**
66. **When MPP is maximum __________ APP< MPP.**
67. **When TPP is decreasing __________ MPP is decreasing.**
68. **APP< MPP __________ Ist stage of production.**
69. **When APP is decreasing __________ MPP< APP.**
70. **Elasticity is negative __________ IIIrd stage of production.**

Finance and Cooperation:

71. **First cooperative Act cooperative credit societies Act 1904 was passed on __________ 1904.**
72. **GOI appointed Sir Edward Mac lagan to probe into performance of the societies. He submitted report in 1915. Both 1912 Act and this report introduced Cooperative planning process in India in __________ 1914.**
73. **Montage Chelmsford reforms by GOI Act, 1919 made cooperative a state subject __________ 1919.**
74. **First land mortgage bank started in Jhang (Punjab) __________ 1920.**
75. **Bombay province enacted Bombay cooperative societies Act __________ 1925.**
76. **Multi-unit cooperative societies Act 1942.**
77. **Lead bank scheme started based on recommendation of Shri F. K. F. Nariman Committee appointed by RBI to look into suggestion of D.R. Gadgil study group which suggested 'Area Approach'. __________ 1969.**
78. **Nationalization of 14 major commercial bank having deposit not <50 crore under Banking companies __________ 19th – July- 1969.**
79. **DIR (differential rate of interest) scheme started based on B. K. Hazare committee recommendation appointed by RBI __________ 1975.**

80. Nationalization of 6 major commercial bank having deposit not < 200 crore__________ **15th-April - 1980.**

81. NABARD (National Bank for Agricultural and rural development) established with merger of ARDC (Agricultural refinance and development corporation), ACD of RBI and RPCC (Rural planning credit cell) of RBI into NABARD__________ **12th-July- 1982.**

82. Indian Companies Act was enacted in **1956.**

83. Which commercial bank became the first in providing mobile ATM services **ICICI Bank.**

84. Wadi programme started by __________ **NABARD.**

85. The headquarter of SIDBI is located in __________ **Lucknow.**

86. Which state in India is the only state to provide Industry status to Agriculture **Maharashtra.**

87. The apex organization of industrial finance in India __________ **IDBI.**

88. What is liquid gold __________ **Petrol.**

89. All India Rural Credit Survey Committee was constituted by __________ **RBI.**

90. The word 'Credit' is derived from __________ **Latin word Credo.**

91. Standard money is also referred to as __________ **Full bodied money.**

92. The symbol used for a graded agricultural produce is__________ **AGMARK.**

93. A large final payment is made at the end of the **loan period** following a series of smaller.

94. Partial payments in case of __________ repayment plan __________ **Lumsum.**

95. Time lag between investment & returns from the agricultural projects is termed as __________ **Gestation period.**

96. Farm clinics is the main feature of __________ **SBI.**

97. In case of DIR scheme, loans are advanced @ of **4%**/annum.

98. A wholesaler assuming the function of retailing is a case of __________ **Forward integration.**

99. Pricing efficiency is also called as **Allocative** efficiency.

100. **Market efficiency** is essentially a degree of market performance.

101. It indicates earning power of project over a period of time __________ **IIR (internal rate of return).**

102. Cash on hand is an example for __________ **Current asset.**

Marketing

103. Market is a place where __________ **Buying and Selling take place.**

104. Who coined the term 'Monopsony' __________ **Mrs. Joan Robinson.**

105. For which of the market there exists no supply curve __________ **Monopoly.**

106. First regulated market in India was established at __________ **Karanjia.**

107. Homogeneous product is characteristic of __________ market **Pure Oligopoly.**

108. Who recommend the MSP for the crop produce __________ **CACP.**

109. National markets are found for __________ **Durable goods.**

110. Regional markets are found for __________ **Food grains.**

111. Short period markets are found for __________ **Perishable goods.**

112. Which market is permanent in nature __________ **Secular market.**

113. Food grain markets, vegetable markets, wool markets are the example of __________ **Specialized market.**

114. A market in which there are more than two but still a few sellers of a commodity is termed as __________ **An oligopoly market.**

115. Central Agmark Lab is located at __________ **Nagpur.**

116. The Agricultural Produce (Grading and Marketing) Act was passed in the year **1937.**

117. The marketed surplus is __________ to the marketable surplus for perishable commodities __________ **Equal.**

118. DMI headquarter is located at __________ **Faridabad.**

119. Marketable surplus is given by __________ **MS = P – C.**

120. WTO came in to effect from __________ **1995.**

121. The Codex Alimentations Commission (CAC) was established in __________ **1963.**

122. __________ is not a function of marketing **Assembling.**

123. __________ is considered as market lifeblood **Market information.**

124. What is "support price" for an agriculture commodity__________ **The minimum price at which the Government is prepared to buy it.**
125. Regulated market ensures __________ **Remunerative price.**
126. The Agricultural Price Commission was set up in the year __________ **1965.**

Important books related to economics:

127. Principles of Economics __________ **Alfred Marshall (1890).**
128. The economics of Imperfect competition __________ **Mrs. Joan Robinson.**
129. Wealth of nation __________ **Adam Smith (1776).**
130. Principles of political economy, Question on some unsettled questions on political economy __________ **J. S. Mill.**
131. An essay in nature and significance of economic science __________ **Lionel Robbinson (1931).**
132. Theory of wages __________ **J. R. Hicks.**
133. Capital and Growth __________ **J. R. Hicks.**

List of important institutes or committees along with their place and year of establishment

Name of establishment	Place	Date/year of
134. Cotton corporation of India (CCI)	Mumbai	1970
135. Agricultural and processed food products export development authority (APEDA)	New Delhi	13-02-1986
136. National dairy development board (NDDB)	Anand	1965
137. Coffee board	Bangalore	21-12-1940
138. Rubber board	Kottayam	1947
139. Tobacco board	Guntur	01-01-1976
140. Spice board	Cochin / Kochi	1987
141. National Agricultural co-operative marketing federation of India (NAFED)	New Delhi	October 1958

142.	National cooperative New development corporation	Delhi	1963
143.	National institute of agricultural marketing (NIAM)	Jaipur	08-08-1988
144.	Indian society of agricultural marketing (ISAM)	Nagpur	1986
145.	Bureau of Indian standards (BIS)	New Delhi	01-04-1987
146.	Indian standards institute (ISI)	New Delhi	1947
147.	Food corporation of India (FCI)	New Delhi	01-01-1965
148.	Directorate of marketing	Faridabad	01-01-1935
149.	Rythu Bazars	Andhra Pradesh	Jan -20-1999
150.	Apani Mandi / Kisan mandi	Punjab	1987 -1988
151.	Indian grain storage	Hapur (UP)	1958

OOO

CHAPTER 10

Agricultural Extension

1. The Goal of Extension Education is an **all-round development of rural people.**
2. Extension first used in a practical way by **James Stewart** of Trinity College, Cambridge University in 1867-68.
3. James Stuart -**"Father of University Extension"**.
4. In a formal way, **Cambridge University in 1873, London University in 1876 and Oxford University in 1878** developed extension system.
5. In 1880 James Stuart work came to known as **"Extension movement"**.
6. The word Extension was used in the form of extension education **in U.S.A.**
7. Philadelphia Committee in 1785- suggested **'Out of school education'** should be given to them.
8. **Smith Lever Act (1914)** was passed for integration of cooperative extension service.
9. Out of school education was organized according to ________. **Smith Leaver Act (1914).**
10. Sriniketan project was closely guided by ________ **Leonard Elmhirst.**
11. Concept of Village Level Worker (VLW) was initiated in ________. **Etawah project.**
12. Concept of Village guide was introduced by ________ **F.L. Brayne in gurgaon project.**
13. Marthan dam project was started with the help of ________ . **YMCA in travencore.**

14. The person associated with Firka development scheme ________. **T. Prakasam.**
15. Chairman of Grow More Food Enquiry Commitee **V.T. Krishnamachari.**
16. Each community development project has an operational area about ________. **400-500 square miles**
17. Number of community development projects launched during **2nd October 1952 is 55.**
18. Community development project was started on the basis of **GMF enquiry report.**
19. The committee which recommended the establishment of panchayat raj system in **1957 was Balwantroy Mehta committee.**
20. Ashok Mehta committee recommended the establishment of ________. **Mandalpanchayat.**
21. SFDA, MFAL projects initiated during which Five year plan ________. **Fourth**
22. SFDA, MFAL is associated with ________. **Venkatapaiah committee**
23. Intensive Agricultural District Programme (IADP) was initiated according to which committee report ________. **Ford foundation.**
24. The report submitted by ford foundation is **India's food crisis and steps to meet it.**
25. IADP was started on pilot basis in which district of **Andhra Pradesh West Godavari.**
26. IADP was developed on the basis of **10 point programme.**
27. Panchayat raj system was adopted first in **Rajasthan and Andhra Pradesh.**
28. **Ashok Mehta** committee recommended the establishment of Mandalpanchayat.
29. **ATMA** programme has its main objective of integrating research, extension and all the stakeholders at district level to support farmers need and interest.
30. ATMA governing board (GB) is an **Advisory body.**
31. ATMA governing board generally consists of **16 members.**
32. ATMA management committee is headed by **project director.**
33. Members of GB of ATMA are appointed for a period of **2 years.**

34. ATMA management committee generally consists of **12 members.**
35. Farmers information and advisory centers under ATMA operates at **block level.**
36. ATIC has the prime objective of providing technological products, services and information to the farming community at the institutional level through **"Single window approach."**
37. The extension component of NATP is **innovation in technology dissemination.**
38. The body which works for policy making and decides on the activities of ATIC is **SITE management committee.**
39. Zonal Research and Extension Action committee (ZREAC) meets **thrice a year.**
40. **Farming Systems Research Extension (FSRE)** is a multidisciplinary, participatory methodology for technology development that merges research and extension efforts.
41. **Farmer led extension** is the process through which individuals or groups within a given locality discover or develop and apply improved ways of managing the available resources, building on and expanding of their indigenous knowledge.
42. **Market Led Extension**is the agriculture and economics coupled with extension is the perfect blend for reaching at the door steps of common man with the help of technology.
43. **Standardization** is concerned with the establishment and maintenance of uniform measurements of produce quality and / or quantity.
44. The process of collecting, interpreting, and disseminating information relevant to marketing decisions is known as **market intelligence.**
45. **Farmer-first model** envisages the supply and demand of innovations as a circular process beginning and ending with farmers, rather than a linear process of beginning with scientists and ending with farmers.
46. **Gender equality** means that there is no discrimination on grounds of a person's sex in the allocation of resources or benefits, or in the access to services.
47. A second emerging model is **farmer-first-and-last (FFL)**. This starts and ends with the farm family and the farming system.

48. **Gatekeeping theory** was first instituted by social psychologist **Kurt Lewin** in 1943.
49. **Ferdinand Tonnies** view of- gemeinschaft and gesellschaft types of society.
50. Agricultural Resources Information Sysem (AgRIS) - **A Needed Domestic Strategy for Sustainable Agricultural Production** and **Sustainable Rural Livelihoods in India.**
51. Research commonly refers to **"Search of Knowledge".**
52. For experimental research **"Internal validity is high, External validity is low."**
53. For non- experimental research **"Internal validity is low, External validity is high".**
54. Observation of "behavior in Natural settings" **–ethnographic study.**
55. Survey research requires the researcher to be a **"trained personal, having social intelligence, manipulative skills, and research insight".**
56. Case study tends to examine a **"social unit as a whole"**
57. **"Research Methods"** indicate an orderly / sequential procedure in research"
58. **"Research methodology"** indicates the science and philosophy behind research.
59. Collection of data and information from human beings takes place in a **"stimulus-response situation"**
60. "**Constructs"** plays a significant role in **"theory building".**
61. **"Control variables"** are those which may affect the relationship between the IDV and DV.
62. The H_0 and H_1 are formulated **"before"** the sample collection work.
63. Reject H_0 when it is true – **"Type I error".**
64. Accept H_0 when it is false – **"Type II error".**
65. Theory helps the scientist to attain **"greater objectivity".**
66. Nature of theory concerned of **"observed natural phenomena".**
67. **Hypothesis** is a conjectural statement **"if such –and – such occurs, than so-and-so –results".**
68. Hypothesis carries clear implication of testing the **relations between variable.**

69. **Do such-and such in so-and - so a manner"** is a **"operational definition"** of research process.

70. **"Cross breaks"** show the elementary way of relation between two variables.

71. Determining the relation between two groups based on **" Cartesian product"** of the two sets.

72. Social scientist commonly calculate **indices of relation,** usually called **"coeficient of correlation"**

73. The Squared coefficient of correlation is called **"coefficient of determination"**

74. **Modern Statistical theory** are based on **"random sampling and its definition.**

75. Thematic Apperception Test **(TAT)** developed by **" Henry Murry"**.

76. Scientific research does not give a place for **"Metaphysical examination"** (Note: Metaphysical means : Highly abstract and overly theoretical, without material form or substance).

77. Reasoning process where in the researcher collect observation and formulate hypothesis is based on **–"Inductive reasoning"**.

78. "α" **coefficient of variance** indicates **Internal consistency** of the measuring instrument.

79. All valid measurements are **"reliable"**, but all reliable measurements are **not valid.**

80. The property of a test that gives us the same score when conn:ected by different people -**Objectivity.**

81. Lower scale value in thurstone & clave scale indicates **Greater Positive attitude.**

82. Key of accurate sampling is **Randomness.**

83. Equality of two regression coefficient is tested with **ANOVA**.

84. Paper-pencil test is otherwise called **– Verbal test**.

85. Survey researcher should study the **relation** among psychological or sociological variables.

86. In Survey research **Panel techniques** enables the researcher **to study changes in behaviors and attitudes.**

87. Theory of convergent, divergent thinking **– J. Guilford**.

88. **t-test** – comparison between two mean (if σ is unknown).

89. **Z-test** – comparison between two mean (if σ is known).

90. **F-test** –comparison between two variance.
91. Socio –economic status scale- **Pareek & Trivedi** .
92. Social science index is published from **"NEWYORK"**.
93. The logic behind hypothesis testing is **"Deductive logic"**.
94. To penetrate deep into the inner part of the group- we go for **"participatory Observation"** of data collection.
95. Assignment of values to alternatives is called -**"Judgement"**.
96. Measurement in Social science is concerned with **properties of objects.**
97. **Intelligence, aptitude, personality** measures strictly speaking falls under –**"Ordinal level" (continuous variables i.e. rank ordered)**.
98. Level of measurement for **categorical variables- "Nominal" (cannot be rank ordered)**.
99. Lowest level in addition & subtraction based scale – **Interval scale.**
100. First psychological test constructed was –**Binet-Simon intelligence scale.**
101. Reflective, transitive & symmetrical relationship is the characteristic feature of - **Nominal scale.**
102. Rorscharch projection tests forms an example for – **Association test**.
103. Q methodology was put forwarded by - **William Stephenson**.
104. Low error variance results in - **High validity and high reliability**.
105. **Chapin & Greenwood** – Ex-post facto research.
106. **Charles Pierce** – "Four methods of knowing".
107. Picture projective measure of cohesiveness **– L. Libo.**
108. **Factor analysis** is a method for determining the number and nature of the underlying variables among larger numbers of measures.
109. **Q-Sort** to allow the participants to act upon the Q samples.
110. Q Methodology provides a formulation for the systematic study of **Subjectivity**.
111. **Q-Sort** usually contains between **60-120 cards.**
112. In sociometry secondary choice star are called **Aristotle leader**.
113. The case-study method was first **introduced into social science by Frederic Le Play** in 1829.

114. **"Content Analysis"** is a method of studying and analyzing communications in a systematic, objective and quantitative manner to measure variable.
115. **"Adorno Scale"** is the E (ethnocentrism) scale.
116. **'Interview'** is the most ubiquitous method of obtaining information from people.
117. Criterion related validity is otherwise called as **"predictive validity"**
118. Reliability is the **"Accuracy or Precision"** of a measuring instrument.
119. **Reliability** can be defined as the **relative absence of errors of measurement** in a measuring instrument.
120. Improvement of Reliability –**"MAXMINCON principle"**.
121. **Regression coefficient** is **independent of Change of origin**, but **dependent on scale value.**
122. Correlation coefficient **independent of both Change of origin,** and **scale value.**
123. The measurement procedure in social research is **"isomorphic to reality"**.
124. Field experiment is a research study in a **"realistic situation"**
125. In Agricultural Extension, good action research is **"Participatory and developmental"**
126. **Complex research design** with **more control groups** often requires more information than a simple group design.
127. **Case Study** aims at deep and detailed study of the unit.
128. **Indexes** are similar to an **inventory** except just consists of one area, a number of dichotomous questions are scored together on the assumption that they have common.
129. **Improving the quality of measurement** by the technique of **"triangulation"** and using a variety of methods.
130. The reliability coefficient tells what proportion of the test variance is **non-error variance.**
131. Psychology originated from ________. **Philosophy**
132. Sociology originated from _______. **Latin, greek**
133. National press day celebrated on ________.**16 November**

134. A belief that technology causes changes in the society______. **Technological determinism**

135. Concept of "social evaluation" first given by ________. **Herbert spencer**

136. Available range of sound waves for human being ______. **20 to 20,000 Hz**

137. Whatever experienced by a person is called ________. **Perception**

138. Need cause a psychological thought or feeling called ________. **desire**

139. The term rurarbinate coined by ________. **Donald Turtle**

140. Informal type of sanction ________. **Ridicule**

141. The process by which the individual learns to confirm to the norms of the group is called ________. **Socialization**

142. Caste system is harmful because it denies ________. **Social mobility**

143. A desire to meet some internalized standard of excellence is called ________. **Need for achievement**

144. Looking glass self a self development concept developed by ________. **Cooley**

145. Who introduced the term "sociology" ________. **August Comte**

146. The book "Indian village" written by ________. **M.N. Srinivas**

147. Primary groups are ________. **Relationship oriented**

148. Secondary groups are ________. **Goal oriented**

149. The group in which individual refers for advises for different aspect- **Reference group.**

150. Father of modern sociology ________. **Emile Durkheim**

151. Indian type of family is characterized by________ **Consanguineous.**

152. Family consists of husband, wife and adopted children i.e. no blood relation ________. **Conjugal family**

153. The family which sets up after marriage is ________. **Family of procreation**

154. Acceptance of e2nd product of cultural give and take is ________. **Acculturation**

155. The process in which person and group compete with each other for resources or to reach a goal is ________. **Competition**

156. Department of rural sociology was established in ________. **1917**
157. Rural sociological magazine formed in ________. **1930**
158. Tightly knit group based on friendship or common interest and having very strong primary group feeling is ________. **Cliques**
159. According to Taylor "the first social group in modern life that approaches self-sufficiency is" ________. **Community**
160. Number of person whose behaviour is polarized about a temporary center of attraction that leads to interaction and inter-related ________. **Collectivity**

Management

161. The acronym 'POSDCORB' symbolizes the ________. **Activities of an administrator**
162. An expression of discontent or dissatisfaction with any aspect of organization is ________. **grievance**
163. The number of subordinates an officer can effectively supervise is known as ________. **span of control**
164. Conferring of specified authority by a higher authority is called ________. **Delegation**
165. Management grid was developed by ________. **Blake and Mouton**
166. The idea of scientific management was introduced by ________. **Frederick Taylor**
167. **Fayol & Weber** - have advocated positional theory of authority.
168. According to whom administration and authority are intimately linked ________. **Weber**
169. The concept of 'POCCC' in management was given by ________. **Fayol**
170. The principle of supervision is inherent in the principle of ________ **Scalar chain**
171. **Non programmed decisions** ________ are used for unstructured, novel and ill-defined situations of a nonrecurring nature.
172. Power vested in an individual is ________ **Authority.**
173. People in the organization who are directly responsible for achieving organizational goals ________. **Line authority**
174. Father of modern management theory ________. **Henry Fayol**

175. According to whom "management is the art of getting things done through people" ________ . **Mary Parker Follet**
176. Contingency approach is also called as ________ . **Situational approach**
177. Total quality management advocated by ________ . **Deming, Juran and Crosby**
178. Measuring and correcting individual and organizational performance to ensure that events conform to plan is ________. **Controlling**
179. The process of selecting, maintaining and developing personnel in position to fulfil the organizational objective is ________. **Staffing**
180. Discharging duties according to spirit of command ________. **Accountability**
181. Whole is greater than its parts ________ . **Synergy**
182. The ability of individuals or groups to induce or influence the beliefs or actions of other persons or groups ________ . **Power**
183. A small group of knowledgeable person to whom some matter is referred to mainly for detail examination and decision is ________. **Committee**
184. Assembly of representatives from local units of parent organization with common interest ________ . **Convention**
185. TOWS matrix developed by ________ . **Wiehrich**
186. Father of social systems approach to organization and management ________ . **Pareto**
187. The elements of administration proposed by ________. **Gullick**
188. Force field theory developed by ________ . **Kurt Lewin**
189. Contingency approach leadership developed by ________. **Fiedler**
190. Path goal theory of leadership developed by ________ . **Rober House**
191. **Policy** ________ guides to thinking rather than to action.

Training

192. Sensitivity training normally needs a period of ________. **5-10 days**
193. The training method that gets its name from a particular behaviour of honey bee ________. **Buzz session**
194. Tainer-trainee ratio is maximum ________. **Lecture**

195. Trainer-trainee ratio minimum in ________. **Brainstorming**
196. The focus of training and learning are respectively on________. **Event and self**
197. Brainstorming Comprises of ________. **Red and green sessions**
198. Sensitivity training also known as ________ **group, laboratory method, encounter method, Daignostic method.**
199. Simulation method is also known as ________ . **Controlled experience method**
200. A particular technique which gives an insight into the complex behavioural patterns of individuals ________. **Role playing**
201. The extension worker trains the farmer at their farm situation in ________. **Peripatatic training**
202. **Learning** ________ the process by which sensations are perceived, meanings enlarged, decisions made and action taken.
203. Correct sequence of seven step model of ELC is **Climate setting, leaners objective, experience process, generalisation, application, closure.**
204. **Job analysis** is a process of gathering information about functions and operations involved in the performance of a job.
205. **Training need assessment** is a systematic process for establishing priorities and making decisions regarding programme planning, development, and operations.
206. **Managerial skills** include supervision techniques, human relations and conceptual skills.
207. **Training plan** is a document for manpower development prepared on the basis of assessed training needs of the personnel.
208. The behaviour domain that gets the highest emphasis in training and education are **Affective domain and cognitive domain.**
209. **Experimential learning** is the process of transforming experience into knowledge as a basis of action.
210. **Empirical methods of assessing training needs** job analysis, performance evaluation, checklist or questionnaire method, tests, critical incident techniques, forced choice technique.
211. The most important form of group discussion and is more formal in nature is ________. **Seminar**
212. The discussion period that may follow after lecture, is ________. **Forum**

213. **Co-training** is a method of instruction in which two or more trainers plan and conduct training together.
214. **MIDDA model of training developed by** _______ Chamala and cogenhourt **(Model for innovation development diffusion adoption).**
215. Number of sub-systems in MIDDA model-3 **subsystems (R&D, diffusion, Formal system).**
216. Goal free model of training evaluation developed by _______ . **Scriven**
217. Broken square exercise developed by _______.**Pfeffier and Jones**
218. Motivation cycle developed by _______. **Morgan and King**
219. The concept of training is considered very loosed & less precise in _______. **Discrete approach**
220. Microlab technique developed by _______. **NTL, Bethel naine, USA**
221. Personal efficacy, goals, locus of control, achievement imaginary, motive strength measured by _______. **TAT technique**
222. Conceptualization of e-values measured by _______. **Tower building technique**
223. Ring-toss exercise developed by _______. **Rosen D. Andrade**
224. Unfreezing and sharing exercise assessed by _______. **Microlab**
225. The fulcrum/main basis of ELC is _______. **Process**

Fundamentals of entrepreneurship

226. Term entrepreneur had its origin in **French, to undertake or to do something.**
227. Entrepreneurship Development Institute of India (EDII) set up in _______. **1983, HQ:Ahmedabad**
228. EDII sponsored by _______. **IDBI, ICICI, IFCI,SBI**
229. Term entrepreneur first used by _______. **Richard Cantillon, (1755)**
230. Richard Cantillon used first Entrepreneur to refer _______. **Economic activities**
231. Entrepreneurship is a _______. **Process**
232. Entrepreneur is a _______. **Person**
233. Enterprise is a _______. **business unit**

234. Intrapreneur term coined by ________. **Gifford Pinchot III**
235. Persons who resigned from their well paid executive positions/ persons left big corporations to launch their own enterprise are called ________. **Intrapreneur**
236. Entrepreneurship is doing new things, or doing things that are already known in a new way given by ________. **Satpathy**
237. Main motivational force for an entrepreneur, ________. **achievement**
238. Entrepreneurs are **moderate risk takers.**
239. Entrepreneur is an arbitrageur according to **Kirzner.**
240. Entrepreneurship is neither a science nor an art, but it is a process by ________. **Peter Drucker**
241. Entrepreneur as social parasite regarded by ________. **Karl Marx**
242. First to introduce dynamic concept of the entrepreneurial function ________. **Schumpeter**
243. Risk taking is the prime characteristic feature of the entrepreneurship as regarded by ________. **F.B. Hawley**
244. Risk, uncertainity and profit as the prime characteristics of entrepreneurs given by ________. **F.H. Knight**
245. An entrepreneur is someone who exercises some control over the means of productionand produces more than what he can consume in order to sell it for individual income by ________. **Economic activities**
246. **J.B. Say** ________ An Entrepreneur is the economic agent who unites all means of production, land of one, the labour of another and the capital of yet another and thus produces a product.
247. An entrepreneur as an Innovator who carry out new combinations to initiate the process of economic development - **Schumpeter.**
248. An entrepreneur is the agent who buys factor services at certain prices with a view to selling its product at uncertain prices in future by **Richard Cantillon.**
249. **Leon Walrus** ________ treated entrepreneur as the fourth factor of production.
250. Entrepreneur as a change agent according to ________. **Frank Young**

Diffusion of Innovations

251. **Observability** _______ is the degree to which the results of an innovation are visible/observable/demonstrable and communicable to farmers.

252. **Complexity** _______ is the degree to which an innovation is perceived as relatively difficult to understand and use.

253. **Replacement discontinuance** _______ is a decision to reject an idea in order to adopt a better idea that supersedes.

254. At **persuasion stage** the individual becomes more psychologically involved with the innovation.

255. **Antecedents** are those variables present in the situation prior to the introduction of an innovation.

256. **Cultural compatability** refers to the degree to which an innovation is consistent with the existing values, beliefs, habits and tradition in a social system.

257. **Hypodermic needle model**, in mass media had direct influence on audience.

258. The hypodermic needle model proposed by _______. **Katz and Lazarsfield**

259. The theory strength of weak ties was proposed by _______. **Granovetter**

260. The pressure from interpersonal networks to adopt an innovation when more members starts adopting is called _______. **Diffusion effect**

261. The perceived attribute of an innovation that reduces the risk component _______. **Trialability**

262. An uncomfortable stage of mind that an individual seeks to reduce or eliminate is _______. **Dissonance**

263. The social structure acts to impede or facilitate the rate of diffusion and adoption of new ideas through _______. **system effect**

264. The situation wherein potential adopters are assumed as blank slates who lack a relevant experience with which to associate the new idea _______. **Empty vessel fallacy**

265. A decision to reject an idea as a result of dissatisfaction with its performance _______. **Disenchantment discontinuance**

266. The small and often subversive units within a large organization that pioneer in creating technology _______. **Skunkworks**

267. The change from an innovation that neither intended nor recognized by members of a social system ________ . **Unanticipated consequences**

268. **Wilkening** ________ was the first to report to report that adoption process involves decision.

269. The degree to which an innovation is changed or modified by a user in the process of adoption or implementation is ________. **Reinvention**

270. **Digital divide**--indicates those individuals who are advantaged versus those who are relatively disadvantaged by the internet and helps illuminate our understanding of inequality in the consequences of innovations.

271. In S-shaped diffusion curve take-off occurs at **10-20%** about adoption, when interpersonal networks become activated so that a critical mass of adopters being using an innovation.

272. **Technology** ________ is a design for instrumental action that reduces the uncertainity in the cause-effect relationships involved in achieving a desired outcome.

273. **Invisible college** ________ is an informal network of researchers who form around an intellectual paradigm to study a common topic.

274. Reconceptualization topic developed by ________ . **Khun**

275. The concept developed by George simmel ________ . **Stranger**

276. **Invention** ________ is the process by which a new idea is discovered or created.

277. The person associated with Law of imitation ________ . **Gabriel tarde**

278. Which concept leads to development of social distance, heterophily, cosmopoliteness ________ . **Stranger**

279. **Tipping point** ________ . indicates small change in individual against harmful effects, it create big change positivity by communication networks.

280. Knowledge to confirmation indicates ________ . **ID/ Gestation period.**

Communication

281. On an average how much time a person spend communicating verbally ________ . **70%**

282. Types of skills pointed out by berlo are ________. **Encoding & decoding**
283. Message distortion explained by ________. **Kirk**
284. Dimensions of message given by ________. **Berlo**
285. Encoding includes ________. **Writing & speaking**
286. Decoding includes ________. **Reading & listening**
287. Amount of information that can be omitted or added in a noiseless channel ________. **Redundancy**
288. **Message** is the actual physical product of the source encoder.
289. **Cybernetics** is the science of communication.
290. **Technical barriers** are the problem of accuracy of information transmitted from source to receiver.
291. Festinger described purpose of communication as ________. **consumatory & instrumental**
292. Write three phases of communication ________. **expression, interpretation & response**
293. Hovland persuasion model is ________. **receiver- listener oriented**
294. Communication I.Q proposed by ________. **Banja**
295. Selecting information while ignoring the distraction is called **selective encoding.**
296. Three levels of communication are ________. **exploratory, convention, participatory**
297. According to laswell,wright and klapper the aims and functions of mass communication are ________. **surveillance, prescription, transmission, interpretation**
298. Information theory developed by ________. **Shannon Weaver**
299. **Gerbener model** is concerned with perception and production of messages.
300. Concept of feedback was given by ________. **Berlo**
301. Measure of degree of randomness or choice is ________. **Entropy**
302. **Communication competency** is the ability to process information cognitively and ability to explain and predict human behaviour.
303. The person associated with political communication, propaganda, political symbolism ________. **Lasswell**

304. STSRD model given by ________ . **Shanon and Weaver**
305. CMCTAR model given by ________ . **Leagans**
306. According to whom human communication systems as open systems ________ . **Gerbner**
307. Learning occurs through various senses shown by ________ . **Heas and Packer**
308. While, communicating verbally, message is distorted upto ________. **30%**
309. The ability to see things from other person's point of view is called **Empathy**.
310. Faithful performance of communication process by all its elements such as communicator, message, channel, and receiver is known as ________ . **Communication fidelity**

OOO

CHAPTER 11

Agricultural Statistics

1. Formula of coefficient of variance is - **SD/Mean *100.**
2. t test is generally used, when sample size is < 30.
3. The correlation coefficient lies between - **-1 to + 1.**
4. The experimental design, which simultaneously control the variation in two directions is known as **LSD.**
5. The basic principles of field experimentation - **1. Replication 2. Randomized 3. Local control.**
6. The technique of analysis of variance was developed by - **R.A. Fisher.**
7. If fertility gradient of land is in two directions , the experimental design will be - **Latin Square design.**
8. The most commonly used measure of control tendency is - **Arithmetic mean.**
9. Cumulative frequency is also known as – **Ogive.**
10. The term 'Kurtosis' was given by - **Karl Pearon (1906).**
11. Laplace used the normal distribution in - **Analysis of errors of experiments.**
12. The important method of least squares was introduced by - **Legendre in 1805.**
13. The name "Normal distribution" was coined independently by - **Peirce Galton and Lexix, (1875).**
14. Normal distribution is also known as - **Law of error, law of facility of error, lapace's second law, gaussion law etc.**
15. Generalized mean is also known as - **Power mean or Holder mean.**

16. The standard deviation is widely used measure of - **Variability or disperson.**
17. The term standard deviation was first used by - **Karl Pearson in 1894.**
18. If mean, median and mode are equal, then skewness will be – **zero.**
19. Correlations are useful because they can indicate a - **Predictive relationship.**
20. Which ANNOVA is used when the experimenter want to study the effects of two or more treatment variables - **Factorial ANNOVA.**
21. Which ANNOVA is used when the subjects are subjects to repeated measures, in which the same subjects are used for each treatment - **Two- Way Anova.**
22. In field experiments the commonly used design is – **RBD.**
23. For chi- square test, the minimum sample size should be – **50.**
24. Define **biometry** - It deals with observation with living thing.
25. Mean, median and mode are equal in - **normal distribution.**
26. Most frequent number in a data set, is known as - **mode.**
27. The well-known example of skewed distribution is - **Personal wealth.**
28. Harmonic mean is the - **Reciprocal of arithmetic mean.**
29. The term z - test is often used to refer specifically to the - **one sample location test.**
30. t Test is a - **Parametric test.**
31. The goodness of fit test is also known as - **kolmogorov - smirnov test.**
32. The number of independent pieces of information that go into the estimate of a parameter is called - **Degree of freedom.**
33. The typical symbol for degree of freedom is - **d.f.**
34. F-test was coined by - George W. Snedecor, in honour of sir **Ronald A. Fisher.**

OOO